Agricultural Value Chains - Some Selected Issues

Edited by John Stanton and Rosa Caiazza

Published in London, United Kingdom

Agricultural Value Chains - Some Selected Issues
http://dx.doi.org/10.5772/intechopen.104239
Edited by John Stanton and Rosa Caiazza

Contributors
Agustinus N. Kairupan, Jantje G. Kindangen, Gabriel H. Joseph, Ronald T.P. Hutapea, Ibrahim Erik Malia, Paulus C.Paat, Derek Polakitan, August Polakitan, Jefny B. Markus Rawung, Meivie Lintang, Joula O.M. Sondakh, Payung Layuk, Olvie Grietjie Tandi, Herlina N.Salamba, Nelson Kario, Jonathan Anugrah Lase, Rindengan Barlina, Natalie Beinisch, Azwihangwisi E. Nesamvuni, Tjale Clopus Mahopo, Noxolo Cebisa Nesamvuni, Johan van Niekerk, Paul L. Woomer, Rachel M. Zozo, Sabra Lewis, Dries Roobroeck, Chris Angelo Dominic, Predrag Rajsic, Alfons Weersink, Richard Gray, Istvan Rajcan, John Stanton, Rosa Caiazza, Ingrid Fromm, Mengistie Mossie Birhanu

First published in London, United Kingdom, 2023 by IntechOpen
IntechOpen is the global imprint of INTECHOPEN LIMITED, registered in England and Wales, registration number: 11086078, 5 Princes Gate Court, London, SW7 2QJ, United Kingdom
Printed in Croatia

British Library Cataloguing-in-Publication Data
A catalogue record for this book is available from the British Library

Additional hard and PDF copies can be obtained from orders@intechopen.com

Agricultural Value Chains - Some Selected Issues
Edited by John Stanton and Rosa Caiazza
p. cm.

This title is part of the Sustainable Development Book Series, Volume 7
Topic: Sustainable Economy and Fair Society
Series Editor: Usha Iyer-Raniga
Topic Editor: Antonella Petrillo

Print ISBN 978-1-83768-511-0
Online ISBN 978-1-83768-512-7
eBook (PDF) ISBN 978-1-83768-513-4
ISSN 2753-6580

IntechOpen Book Series
Sustainable Development
Volume 7

Aims and Scope of the Series

Transforming our World: the 2030 Agenda for Sustainable Development endorsed by United Nations and 193 Member States, came into effect on Jan 1, 2016, to guide decision making and actions to the year 2030 and beyond. Central to this Agenda are 17 Goals, 169 associated targets and over 230 indicators that are reviewed annually. The vision envisaged in the implementation of the SDGs is centered on the five Ps: People, Planet, Prosperity, Peace and Partnership. This call for renewed focused efforts ensure we have a safe and healthy planet for current and future generations.

This Series focuses on covering research and applied research involving the five Ps through the following topics:

1. Sustainable Economy and Fair Society that relates to SDG 1 on No Poverty, SDG 2 on Zero Hunger, SDG 8 on Decent Work and Economic Growth, SDG 10 on Reduced Inequalities, SDG 12 on Responsible Consumption and Production, and SDG 17 Partnership for the Goals

2. Health and Wellbeing focusing on SDG 3 on Good Health and Wellbeing and SDG 6 on Clean Water and Sanitation

3. Inclusivity and Social Equality involving SDG 4 on Quality Education, SDG 5 on Gender Equality, and SDG 16 on Peace, Justice and Strong Institutions

4. Climate Change and Environmental Sustainability comprising SDG 13 on Climate Action, SDG 14 on Life Below Water, and SDG 15 on Life on Land

5. Urban Planning and Environmental Management embracing SDG 7 on Affordable Clean Energy, SDG 9 on Industry, Innovation and Infrastructure, and SDG 11 on Sustainable Cities and Communities.

The series also seeks to support the use of cross cutting SDGs, as many of the goals listed above, targets and indicators are all interconnected to impact our lives and the decisions we make on a daily basis, making them impossible to tie to a single topic.

Meet the Series Editor

Usha Iyer-Raniga is a professor in the School of Property and Construction Management at RMIT University. Usha co-leads the One Planet Network's Sustainable Buildings and Construction Programme (SBC), a United Nations 10 Year Framework of Programmes on Sustainable Consumption and Production (UN 10FYP SCP) aligned with Sustainable Development Goal 12. The work also directly impacts SDG 11 on Sustainable Cities and Communities. She completed her undergraduate degree as an architect before obtaining her Masters degree from Canada and her Doctorate in Australia. Usha has been a keynote speaker as well as an invited speaker at national and international conferences, seminars and workshops. Her teaching experience includes teaching in Asian countries. She has advised Austrade, APEC, national, state and local governments. She serves as a reviewer and a member of the scientific committee for national and international refereed journals and refereed conferences. She is on the editorial board for refereed journals and has worked on Special Issues. Usha has served and continues to serve on the Boards of several not-for-profit organisations and she has also served as panel judge for a number of awards including the Premiers Sustainability Award in Victoria and the International Green Gown Awards. Usha has published over 100 publications, including research and consulting reports. Her publications cover a wide range of scientific and technical research publications that include edited books, book chapters, refereed journals, refereed conference papers and reports for local, state and federal government clients. She has also produced podcasts for various organisations and participated in media interviews. She has received state, national and international funding worth over USD $25 million. Usha has been awarded the Quarterly Franklin Membership by London Journals Press (UK). Her biography has been included in the Marquis Who's Who in the World® 2018, 2016 (33rd Edition), along with approximately 55,000 of the most accomplished men and women from around the world, including luminaries as U.N. Secretary-General Ban Ki-moon. In 2017, Usha was awarded the Marquis Who's Who Lifetime Achiever Award.

Meet the Volume Editors

John L. Stanton has a Ph.D. in Quantitative Methods and Marketing from Syracuse University, USA. He has been in the food industry for about 40 years. He is currently a professor in the Food Marketing Department, Saint Joseph's University, USA. Dr. Stanton was elected to the European Retail Academy Hall of Fame and the Private Label Manufacturers Hall of Fame. He is the recipient of a Lifetime Achievement Award from the Promotion Optimization Institute. In addition to his career in academia, Dr. Stanton has also worked in the food industry. He has consulted for many nationally known food companies including Campbell Soup Company, Procter & Gamble, Acme, Kroger, Pepsi, Frito Lay, Florida Department of Citrus, Kellogg, and others. Dr. Stanton has more than 100 peer-reviewed articles and 13 books to his credit.

Rosa Caiazza, Ph.D., is a Professor of Management at Parthenope University of Naples, Italy, and a visiting professor at Wharton School of the University of Pennsylvania, USA. She is listed among the top 2% of scientists according to Stanford University, USA. She has been included in the International Council for Small Business's (ICSB) list of 300 of the world's most well-renowned professors of small business and entrepreneurship. She serves as an advisory board member for numerous top-tier academic journals, including the *Academy of Management Perspective, Journal of Management Studies, The Journal of Technology Transfer, Small Business Economics, Trends in Food Science & Technology, Management Decision, Journal of Small Business Management, British Food Journal, Journal of Management Development, Journal of Organizational Change Management*, and many more. Her research and teaching activity focuses on strategy, corporate governance, entrepreneurship, innovation, and operation management. She has published 4 books and more than 100 journal articles and book chapters. She has served as chairman of many international conferences and has won several Emerald Literati Awards for Excellence, including the Outstanding Paper Award.

Contents

Preface

This book focuses on the food supply chain, especially in lesser developed countries where supply chain problems and issues in the agribusiness sector are most dramatic. However, the lessons learned from this book are pertinent to any aspect of the supply chain and to any country whether it is fully developed or not. The chapters cover many aspects of the supply chain or the value chain of agribusiness, ranging from global trends to food production.

In Chapter 1, "Theoretical Underpinnings of Value Chain Analysis", Mengistie Mossie Birhanu sets the stage for the rest of the text by examining the basic concepts of the supply chain. The chapter collects important secondary data to corroborate facts and to understand key issues pertinent to relevant theories. In Chapter 2, "The Importance of Short Food Supply Chain: Examining Logistics Strategies for Competitive Advantage and Sustainability", Chris Dominic writes more generally about the supply chain and the importance of logistic strategies. In Chapter 3, "Trends Shaping the Future of Agrifood", John Stanton and Rosa Caiazza write about trends shaping the future of the agribusiness supply chain. In Chapter 4, "Technology Promotion and Scaling in Support of Commodity Value Chain Development in Africa", Paul L. Woomer et al. discuss technology promotion and scaling in support of commodity value chain development in Africa.

Speaking of Africa, developing countries and supply chains are another key element of this book. It is important to note that in many countries around the world, wet markets or rural food stores dominate the retail market. In some countries, only 14% to 20% of all food sales are from modern markets. In Chapter 5, "Value Chain Analysis of the Street Food Enterprises in the Rural Towns of Vhembe District, Limpopo Province", Tjale C. Mahopo et al. present a value chain analysis of street food enterprises in the rural towns of the Vhembe District, Limpopo Province. While their analysis focuses on two towns, the lessons learned go beyond those two places.

Not all the chapters are about supply chains and retailers. In Chapter 6, "Reducing Inequalities in the Coffee Value Chain: Threats and Opportunities for Small-Scale Farmers in Central America and East Africa", Ingrid Fromm writes about the value chain and small-scale farmers. This topic is very pertinent, as local products are more and more valued, but unless the supply chain can bring these products to market it will not be a successful venture. In Chapter 7, "Value Chain Implementation in Rural-Scale Integrated Coconut Farming System in North Sulawesi Province, Indonesia", Agustinus N. Kairupan et al. discuss issues with farms and their value chain implementation in the rural-scale, integrated coconut farming system in North Sulawesi Province, Indonesia.

Finally, two of the chapters go back to food itself. In Chapter 8, "Are There Adequate Incentives for Research and Innovation in the Plant Breeding Supply Chain?", Predrag Rajsic et al. write about the need for research and innovation in plant breeding in the supply chain of new and innovative products. In Chapter 9, "Circular Business Models and Global Value Chains: The Case of MagProtein", Natalie Beinisch looks back even further to metaproteins with global value chain research focused on the governance of diffuse modes of production and trade that travel in a single direction; however, circular economy business models are by definition closed economic systems in which it is possible for two parties to act as both buyers and sellers.

John Stanton
Saint Joseph's University,
Philadelphia, Pennsylvania, USA

Rosa Caiazza
Parthenope University of Naples,
Naples, Italy

Chapter 1

Theoretical Underpinnings of Value Chain Analysis

Mengistie Mossie Birhanu

Abstract

Value chain related theories evolve over time in response to critique and their own conceptual development. In addition, they are ways of assessing the real world (theoretical frameworks); generating explanations about market development practice, and provide the opportunity for comparison over space and time. The purpose of this chapter is to provide an overview of the theoretical literature on theoretical underpinnings of value chain analysis. The approach used has been a sort of desktop review which involved collection of important secondary data to corroborate facts and to understand key issues pertinent to the theories was fundamental. After a search for literature, the review employed a method known as content analysis. The review result showed that the theories regarding value chain analysis intended, in time, by simplifying the reality or image to comprehend the phenomena with the purpose of their forecasting. In addition, the theories presented an increased concern, irrespective of the place where they were developed, but also, they stirred some critics. Moreover, it is the belief of this book chapter that it is intended to serve as both a summary of the state of the field and an overview reference for users.

Keywords: theory of change, value chain, strategic business management, transaction cost, new paradigm

1. Introduction

For a large part of the world's growing population, the increasing integration of the global economy has provided the opportunity to achieve significant prosperity gains. For developing countries, the globalization of manufacturing has opened up new prospects of upgrading their industrial and service sectors. It also holds the promise of higher incomes, increasingly differentiated final products, and a greater availability of quality goods. Most notably, free trade agreements and other accords have created new export opportunities—mainly for food products—as the demand for variety continues to grow in developed countries. The expansion of agro-industrial operations and connections to export markets has been supported by these market shifts among governments, investors, and farmers in order to increase local food production, employment, company growth, and international trade. This has led to competition among producers to meet export market demands in terms of cost, quality, and delivery times. Consequently, a wide range of companies have evolved to provide goods and services to help agro-industries meet those demands. At the same time, policies, regulations, support services, tax and trade instruments and their associated actors and institutions have also developed to become intrinsic parts of the so-called "value chains."

Such a move, globalization, is changing the environment in which poverty-reduction strategies are being implemented. In this new context, two things are clear: poverty alleviation cannot be sustained without economic growth; and economic growth cannot be sustained in non-competitive industries. Hence, a value chain approach which focuses on industries employing large numbers of the poor and with the potential to become and remain competitive in global markets should be in place. This approach therefore has relevance in a wide array of programs for which poverty reduction and/or wealth creation is either the ends or the means.

Market globalization generally links a firm's ability to remain viable to the competitiveness of the businesses in which it operates. Firms within an industry in a country or region must increasingly compete—even in local markets—with firms and industries from across the globe. To succeed in global markets, entire industries (or value chains) must be able to deliver a product to the consumer more efficiently, with a higher quality and/or in a more unique form than the value chains in competing countries. This tells us that competitiveness at the firm and industry levels is interdependent. In this chapter, readers will learn the theoretical underpinnings, theory of change and value chain analysis nexus, development and operation of agricultural value chains with practical cases from Ethiopia.

The "value chain" concept has evolved, within academics from a variety of "ideological" schools influencing it along the way [1]. The francophone concept of filière and the anglophone concept of commodity chain underpin value chain theory, analysis, and methodologies [2]. Beginning in the late 1990s, the term "commodity chain" was gradually replaced in the literature by "value chain" [3]. Many researchers appear to agree on the concept of value chains, but their explanations and abstractions of empirical results differ [4]. The two key concepts in the study of value chains are the words "value" and "chain." In the analysis of the value chain, value is a synonym for "value-added," [5] whereas "chain" refers to the processes and parties involved (from conception to disposal) in the lifespan of a product [6]. The term "value chain" has been defined in a variety of ways in the literature. Kaplinsky and Morris [7] defines it as "the entire set of activities necessary to take a product or service from conception to various stages of production, consumer delivery, and final disposal after use." This definition recently was adapted to the specific field of agri-food products by [8], which provided the following explanation:

"The entire range of farms and firms, as well as their consecutive coordinated value-adding practices, that yield various raw agricultural items and convert them into specific food products sold to end-users and disposed of after usages in a way that is cost-effective all around, has wide societal benefits and does not perpetually deplete resources" [8].

This concept has been widely used to understand the effects of market relationships and upgrading processes on smallholders in developing economies. The fundamental feature of a value chain is market-focused collaboration: various business enterprises collaborate to produce and market products and services in an efficient and effective manner [9, 10]. The promotion of value chains in the agricultural sector aims to increase product competitiveness in international and domestic markets, as well as to generate more value-added within the region or country [11, 12]. The value chain approach has gained attention in development practice and policy, primarily as a lens for developing poverty-reduction initiatives with private sector participation.

This has resulted in the development of a widely used analytical approach in development fields known as "value chain analysis" [13, 14].

Value chain analysis is a reformulation of the Orthodox theory of trade, which is based on Ricardo's (1817) law of comparative advantage. In comparison to Orthodox trade theory, the value chain approach is more feasible and is now more commonly used to address poverty reduction and food security efforts, primarily by assisting the poor in gaining access to markets [15, 16]. In explaining why, the poor may face trade barriers and how to overcome them, value chain analysis is more useful than traditional theory. This is because Orthodox trade theory assumptions, such as the link between trade and poverty reduction on the one hand and economic growth on the other hand, have never been the main focus. It also fails to deliver feasible intervention strategies for policymakers and practitioners with more reasonable goals: how and when to assist a recognized target group in accessing (or, in better terms, accessing) specific effective value chains. Recognizing these flaws, trade theory is being amended and, in many important ways, is converging with value chain analysis [17, 18]. The sub-sections in 3.1 show the brief discussions on the development of the concept of value chain as a new paradigm.

2. Methods to be applied

In reviewing this book chapter, the approach used has been a sort of desktop review which involved collection of important secondary data to corroborate facts and to understand key issues pertinent to the theories was fundamental. After a search for literature, the review employed a method known as content analysis.

3. Theoretical underpinnings

The following sub-sections show the brief discussions on the development of the concept of value chain as a new paradigm.

3.1 The Filiére concept

The "Filiére" approach was developed in the 1960s to gain a more detailed understanding of the economic processes that underpin agricultural commodity production and supply chains [19]. In Francophone countries, the approach is used to analyze agri-food supply chains. It incorporates various schools of thought, including system analysis, Marxist economics, industrial organization, management science, and neoclassical welfare analysis [2]. It was soon utilized in the research of agriculture in developing countries, emphasizing growing exports of products such as cocoa, cotton, and coffee from French colonies. This theory is then used to shape French industrial policies. However, the "Filiére" idea failed in its analysis of the global world economy because it is a static model with non-changing participants. Rising or falling commodity or information flows, as well as stakeholder increases and decreases, are not taken into account [10]. Overall, the Filière can be used to describe the flow of physical inputs and services in the production of a finished product, and is conceptually similar to the modern value chain concept [20]. In more recent years, filiére research has been influenced by transaction cost economics (TCE), regulation theory, and convention theory [2].

3.2 Strategic business management perspective

Michael Porter's [21] concept of value chains is found in the literature of strategic business management, under the umbrella of supply chain management (SCM). The concept was created to assist managers in determining the value embedded within their company's various supply and support operations, as well as ways to improve that value in order to gain a competitive advantage. According to Porter, value chain analysis is the ultimate technique for evaluating the profits generated at each point of production. Porter distinguished various stages of the supply process (inbound logistics, operations, outbound logistics, marketing and sales, and after-sales service), the transformation of these inputs into outputs (production, logistics, quality, and continual improvement operations), and the support services the firm marshals to accomplish this task [22].

SCM first appeared in the early 1980s as an approach to managing flow of products and enhancing efficiency within a single company [23]. SCM's boundaries grew to include all actions and actors engaged in delivering a product from raw material to consumer over time, and it now encompasses all activities and actors involved in the delivery of a product from raw material to consumer [24]. Overall, the method strives to integrate and coordinate activities and stakeholders across the supply chain in a systemic and strategic manner in order to optimize the supply chain's performance in providing the product at the lowest possible cost [25].

SCM does not have a unifying theory that underpins it. Instead, a variety of current theories from diverse domains have been employed to describe certain elements, meaning that no single theory can adequately cover the scope of SCM [26, 27]. According to [28], the combination of TCS, the resource-based view, and network theory is most beneficial in explaining supply chain structure and management issues. These authors suggest that when the three theories are considered together, they can create a mid-level theory for SCM that they call the "new institutional economics approach." Porter's concept of competitive advantage is also incorporated into SCM. According to Porter, enterprises or chains must offer consumers either cheaper pricing or unique product or service features in order to establish and retain competition [7].

Transaction Cost Theory: The transaction cost theory is classified as a branch of New Institutional Economics (NIE) [29, 30]. Transaction costs are expenses incurred when activities such as information gathering, negotiation, bargaining, enforcing, and monitoring are carried out [31]. The costs of identifying markets and trading partners, as well as the costs of obtaining price and product information, are referred to as information costs. The expenses of physically negotiating, bargaining, and formally drafting the terms of exchange are referred to as negotiation costs. Monitoring and enforcement costs are the expenses of making sure that the transaction's terms, like quality standards or payment methods, are followed by the trade agreements [32].

The business relationships between suppliers and purchasers produce "transaction costs" in addition to the costs of manufacturing and marketing at each phase of the value chain. Market inefficiencies, like low market consistency, lack of grades and specifications, or weaknesses in the business environment, often result from high transaction costs. TCE explores the basis for governance decisions about inter-company organizational relationships. Inter-organizational relationships are agreements that bring together organizations with the goal of producing joint added value [33]. In TCE, the fundamental unit of analysis is transactions among firms [34]. Under the conditions of stakeholders' opportunistic actions, businesses choose a governance form that significantly reduces transaction costs. Through mutual investment, control mechanisms, and complex organizational structures such as contracts, value chain actors safeguard against the risk of opportunity. The majority of small and marginal farmers in underdeveloped nations live in remote areas with inadequate infrastructure, and they frequently miss out on market opportunities due to high transaction costs [35]. According to [36] transaction costs negatively impact market participation, while better market information triggers it. Barrett [37] argues that distance to market is a proxy for transaction costs and has a negative impact on market participation. Overall, the idea of transaction costs aims to clarify what kind of governance system is empirically defined [38].

Network theory: This is another relevant theoretical stream for developing value chain research. It relies on the interdependence of social and economic interactions in (production) networks consisting of multiple vertical and horizontal relationship issues among stakeholders in the supply chain [39, 40]. According to the network framework, businesses are integrated in a dynamic of vertical, horizontal, and business development relationship issues with several other institutions that provide inputs such as credit coordinators, advisory services, and transportation services [41]. The supply chain demonstrates vertical connections among market participants in order to co-produce for a market, whereas network theory combines vertical and horizontal interactions between actors [42]. Network linkages can boost a company's "social capital" by making it easier to obtain information, necessary knowledge, and financial assistance, as well as by empowering information sharing among key stakeholders lowering transaction costs and enhancing access to markets [43]. When the value chain is more described by a network than by a single vertical chain, network analysis is used. To summarize, networks are an important factor in the development of both intervening and upgrading value chains, and they serve as a basis for analyzing and mapping relations and flows among people and organizations [44].

3.3 Political economy perspective

The third important antecedent of value chain study is political economy research, which looks at the causes, nature, and effects of global industrial and technological integration. Wallerstein's world-systems analysis, which emphasizes on the world-system as the unit of social analysis and is focused on understanding the varied implications of the capitalist world economy, provides the foundation for this line of research, which is referred to as "global chains" research. As a result, one of the most important themes is the distribution of power and advantages among people from industrialized and developing countries. Despite the fact that the literature on global chains is broad and multidisciplinary, three important analytical frameworks may be distinguished: global commodity chains, the world economic triangle, and global value chains (GVCs) [45].

Gereffi developed the "Global Commodity Chain (GCC)" approach in the mid-1990s, which is based on Wallerstein's commodity chain [46]. This perspective is used to investigate the origins, nature, and consequences of global technological and industrial integration. The approach builds on the world systems perspective. Gereffi's contribution has supported significant advances in the empirical and normative application of the value chain framework, especially for its emphasis on the power structures embedded in value chain analysis. The approach focuses on the power relations embedded in value chain analysis and connects the concept of value-added chain to global industry organization [47]. Moreover, in 2002, Messner developed the theory of the "world economic triangle" where actors, governance, and regulation systems are influential factors for regions (developing countries) scope of action in the global commodity chains. The six major characteristics in all economic triangles are: group of actors, their interests, trust, power relations, action-orientation, and mind sets [48]. World economic triangle approach focuses on the upgrading of whole regions or clusters through their integration in value chains, which might bring up grading perspectives for regions in developing countries. Therefore, this theory links horizontal (cluster development) and vertical approaches (value chain) [49].

In the mid-1990s, the growing fragmentation of supply chains in the global arena led to the development of the literature. This literature added an explicitly international perspective and centered on worldwide power dynamics and rule-setting processes (governance) across the chain [50]. Humphrey and Schmitz [51] introduced the "Global Value Chain" concept. In 2000, a team of researchers with considerable expertise met to establish a theory of "global value chains" [52] in a series of seminars. Its goal was to construct an exact theoretical model appropriate to real-world scenarios, robust, and applicable [19]. The advantages of Porters' single firm orientation are recognized by this new school of thinking (GVC), while at the same time scrutinizing relations between the various stakeholders. A GVC approach enables social influences to be incorporated and provides a structure that connects constraints, governance structure, and upgrading strategies [2].

The governance structure is a crucial step to understand the nature of coordination and relationship mechanisms that exist between stakeholders in the chain [47]. Governance involves coordination and related positions in finding vibrant cost advantages and allocating key players' roles [7]. Governance's central premise is the realization that encounters between businesses in a value chain reveal organizational forms rather than merely random interactions [7], which implies that governance guarantees that the organization is represented by interactions among entities across a value chain [47]. It is noted that management and coordination often function in literature as interchangeable or synonymous terms [53]. The term governance was used by Williamson [54] in the 1980s to describe the collection of hierarchical frameworks in which a transaction is organized. Based on governance structures, value chains can be classified into producer-driven and buyer-driven value chains [47]. However, in developing countries, where food chains still have many stages, most transactions are carried out with the assistance of middlemen. They may be able to address information asymmetries that are prevalent in developing countries, but they may also take a greater share of the market share in the value chain, limiting upgrading possibilities for farm households [55, 56].

Most empirical findings [57, 58] in developing economies reveal that middlemen take the lion's share of the margin shares and are more influential in the chain. The hegemony of these middlemen in the value chain can be explained in part by the terrestrial dispersion of actors and the scarcity of price information. Because of the chain's extreme geographical dispersion, the actors require an assistant who can act on their behalf in other markets and provide price information [59]. Due to a lack of valuable market information and a lack of well-organized participants in developing countries, participants rely heavily on middlemen. Most of the time, middlemen purposefully create a communication chasm among both buyers and sellers and arbitrate them in their preferred manner. As a result, middlemen are viewed as market impediments by both buyers and sellers. The price is also set by the middlemen. In such cases, neither the producer nor the buyer has the power to determine exact product prices [60].

Middlemen's actions have an impact on customer prices since they often collect and manipulate selling prices. Ignoring middlemen and believing that producers/growers could be tied directly to purchasers without some sort of middle, value-added function is generally ill-informed and contributes toward further market imperfections and decreased competitive advantage. As a result, middle actors in value chains should be enticed with incentive schemes to drive better business operations based on effective competitive environment and compliance with laws and regulations. Munshi [61] argues that middlemen are not only an economic institution but also a social network structure that facilitates trade in developing countries. In addition to the two types of value chains described above, unlike developed countries value chains, this chapter proposes a third type of value chain for developing countries like Ethiopia known as the "middlemen-driven value chain." This means that three types of governance are distinguished based on the coordination role of "governance": those where producers have dominance ("producers-driven value chain"), those where middlemen have hegemony ("middlemen-driven value chain"), and those where buyers have a key role ("buyers-driven value chain").

Long marketing channels benefited middlemen while negatively impacting both producers and consumers. This meant that a long marketing channel was one of the main causes of increased transaction costs as well as crop marketing inefficiency. As a result, the government should devote sufficient attention to improving the imperfect market chain as well as long distribution channel by establishing institutions such as cooperative-unions. Brokers should be formalized into legal venture for the benefit of growers, wholesalers, retailers, and consumers by recognized financial institutions. This situation clearly necessitates fierce government interference. This is because unregulated middlemen are demolishing a considerable amount of value [59, 62]. In general, the governance structure provides information concerning farm households' positions in the value chain as well as the relationships among farmers and buyers.

Another key term in the GVC literature is upgrading, which is seen as a complement to governance. Upgrading is referred to as strategies for adding value, which is the intervention step of value chain analysis [63]. Upgrading is defined by Mitchell [15] as the "means of acquiring the technical, institutional, and market capabilities that enable poor communities to enhance their competitiveness and shift toward higher-value activities." Value chain players are said to upgrade as they gain new skills by creating more value-added commodities or improving existing ones. Upgrading options could be specified by assessing profitability inside the chain and defining its constraints. Improving strategies that include interventions such as improving product quality and moving toward more creative market segments, redesigning the production line or investing in new process upgrade techniques, improving the chain's performance, and introducing innovative features to increase the quality of activities to gain deeper importance throughout chain [64]. In various countries and businesses, empirical research [65] offers information about the benefits of upgrading agriculture enterprises.

Various scholars have identified four upgrading opportunities namely: process upgrading, product upgrading, functional upgrading, and inter-chain upgrading [64]. Process upgrading focuses on enhancing the efficiency of external and internal processes within the value chain. For instance, processes that ensure timely deliveries, collection of quality products, or improved marketing of a product, organizational restructuring, collaborations, and/or capability buildings are ways to achieve process upgrading [7]. Product upgrading refers to improving existing products and/or developing new ones. It is closely linked to process upgrading because changes in products often lead to changes in processes [15]. Functional upgrading is achieved when a firm changes one's position within the chain to add value. An example of this can be farmers who start processing in addition to producing fruits. On the other hand, upgrading of the chain involves moving to a new chain. If participating in one chain is not profitable, farmers may look for other options, since they have a diversified livelihood strategy. But high barriers of entry into new value chains might limit their options [7].

3.4 Theory of change and value chain analysis nexus

Theory of change, which was developed in the 1990s, is a method for designing business strategies that brings together community development partners and addresses proposed way of achieving impact [66]. In the case of smallholder value chains, theory of change was used to develop business modeling techniques aimed specifically at integrating smallholder farmers into market-oriented value chains [67, 68]. It is a good place to start when it comes to understanding and clarifying smallholder commercialization transactions. It demonstrates how well the various interventions and tasks will interact to produce the desired market changes and, ultimately, the planned impacts on smallholder farmers for each value chain. It is based on the idea that by supplying farmers with the appropriate inputs, they will be able to expand their farming production, resulting in higher earnings and therefore more sustainable farmer practices [67]. The main goal of such assistance programs should be to enable smallholder farmers to make appropriate, well-informed decisions about how to manage their agriculture and which markets to pursue [69].

Consider an intervention designed to enhance household food security and welfare as an example (i.e., final impacts). Inputs could include fertilizer, pesticides, seeds, and pest management (**Figure 1**). However, inputs include not only production techniques, but also expertise, such that poor and disadvantaged farm households can use the accessible information for personal gain. Connecting smallholder farmers to well-functioning markets ranging from local markets to structured value chains is critical in long-term rural poverty-reduction strategies. Comprehending how to successfully connect disadvantaged growers to markets, as well as recognizing that markets can profit which types of participants, are important steps in community development [70, 71]. The theory of change depicted in **Figure 1** illustrates a procedure that helps to identify critical steps and components when forming actor partnerships. Some of the strategies to increase profits for farm households include going to invest in upgrading the value chain to meet production and distribution requisites, making investments in wider sustainable livelihood strategies, as well as trying to adapt trade relations and value chain formation for smallholder sourcing.

The intervention logic (theory of change) for a value chain method is depicted in **Figure 1**. These strategies include expanding input access, increasing productivity, enhancing value chain connections, expanding access to markets, and improving supportive framework (an enabling environment). Increased access to productive resources such as agricultural inputs, land, pest management, irrigation, as well as other extension services can help disadvantaged farmers increase their productivity. Similarly, it is critical that they cultivate their relationships with other actors in the chain. As a result, building strong value chain partnerships among farm owners as well as other actors in the chain through coordination, communication, and contracts is as significant as the other strategy elements. Finally, an enabling environment should be developed and enhanced through education and advisory services. These schemes would produce both tangible (higher profits, better access to capital, and

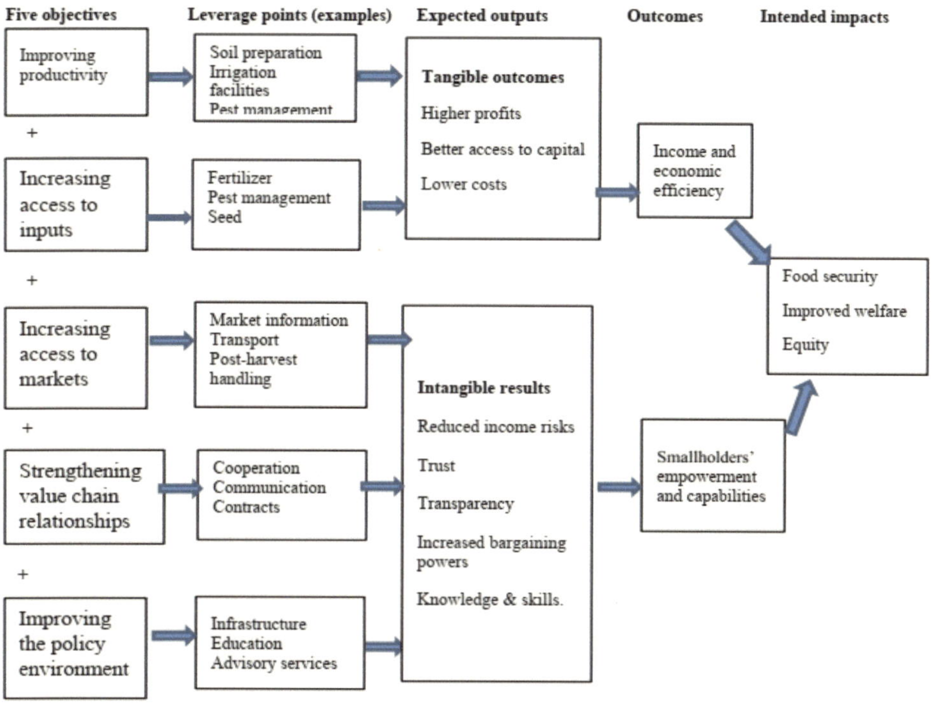

Figure 1.
Value chain theory of change, objectives, outputs, and outcomes. Adopted from [69].

lower costs) and intangible results (reduced income risks, trust, transparency, increased bargaining powers, and knowledge and skills). Smallholders would benefit from food security, as well as a reasonable income and environment. Despite efforts to incorporate value chain expansion into developing nations' poverty-reduction strategies, initiatives are frequently ineffective. One reason for this is a lack of coordination efforts throughout the entire value chain [69, 70].

3.5 Sustainable livelihood framework

Participation in the agri-food value chain knowledge, good health, and the ability to work that allows people to pursue various livelihood choices and achieve their goals. Human capital is a factor of the amount and quality of labor available in a household. These changes depending on the size of the household, skill levels depend on a number of factors at the producer level. There are several perspectives from development economics for understanding what determines smallholders' participation in agri-food value chains, including the sustainable livelihood framework (SLF), which includes five types of assets or capitals: natural, physical, human, financial, and social capitals [72]. Due to the quantities of assets envisaged by such individuals, the capitals explain why individuals in rural areas are able to participate in some activities while others are not. According to [73], varied value chain participation among smallholders could be explained by variable access to assets and services to decrease transaction costs.

The following are the explanations behind these capitals: *Natural capital* refers to the natural resource stocks (e.g., land, trees, water, air, genetic resources, etc.) from which resource flows and environmental services (e.g., nutrient cycling, hydrological cycle, erosion protection, pollutant sinks, and etc.) are produced that are important for livelihoods. *Human capital* includes things like literacy, skills, health, and the ability to work. Human capital is the combination of skills, leadership potential, and health situation, among other factors. Human capital occurs as a livelihood asset in the framework for sustainable livelihoods, that is, as a component or means of achieving livelihood outcomes [74]. *Physical capital* refers to the infrastructure and production goods that are required to support livelihoods and pave the way for livelihood strategies. Infrastructure includes accessible modes of transportation, proximity to towns and the capital city, a marketing location, suitable housing, water and sanitation, and information access. Infrastructure and producer goods are frequently thought to be required to sustain livelihoods [75]. *Social capital:* different social venues can establish formal and informal interactions from which people might get varied possibilities and rewards in their pursuit of a living. It encompasses vertical and horizontal networks and connectedness that strengthen people's trust and ability to collaborate, as well as their access to larger institutions such as political and civic bodies. *Financial capital* is the term used to describe the financial resources that people employ to fulfill their livelihood goals. Savings, credit, and production equipment are all part of it [74].

SLF provides a paradigm for examining how institutions affect smallholder income, asset holdings, consumption, and poverty in direct and indirect ways. Institutions, according to SLF, influence access to assets or resources, affecting livelihood strategies. For example, a household may choose to intensify or diversify its activities. If a household decides to intensify, like in this case of farmers investing in the production and selling of better fruits varieties, financial, social, and physical assets will all play a role. The livelihood strategy chosen has an impact on the outcomes, such as poverty alleviation [76]. The SLF, in general, explains the resources that smallholders require in order to participate in agri-food value chains.

4. Concluding remarks

In this chapter, we have explored the main theoretical concepts, ideas, and critiques arising from the literature on (agricultural) value chain theories. The value chain approach has gained attention in development practice and policy, primarily as a lens for developing poverty-reduction initiatives with private sector participation. This review paper revealed that there has been a revolution in thinking about value chain theories over the past several decades, since 1960s. Many of these value chain theories provide powerful explanations for growth and circumstances of the approach. In conclusion, theories are developed to provide us understand the value chain concepts and approaches over time. The theories presented an increased concern, irrespective of the place where they were developed, but also they stirred some critics. That is, new theories/approaches can develop on the basis of criticism and the rejection of existing frameworks of understanding and they are constantly evolving in parallel with the world they seek to explain.

Conflict of interest

The author declares no conflict of interest.

Additional information

The book chapter: "Theoretical underpinnings of value chain analysis" is part of my thesis work. It is made available online (link attached below). This can make the document more likely to increase the similarity index.

http://213.55.95.56/bitstream/handle/123456789/29358/Mengistie%20Mossie.pdf?sequence=1&isAllowed=y

Author details

Mengistie Mossie Birhanu
College of Agriculture and Environmental Sciences, Bahir Dar University, Bahir Dar, Ethiopia

*Address all correspondence to: mengistiemossie@gmail.com

References

[1] Fonseca CMB, Coelho JC, Soares FB, Correia A. Value chains analysis: overview and context for development. Direct Research Journal of Agricultural and Food Science. 2019;7(12):356-361

[2] Raikes P, Friis Jensen M, Ponte S. Global commodity chain analysis and the French filière approach: comparison and critique. Economy and Society. 2000;**29**(3):390-417

[3] Stoian D, Donovan J. Putting value chain development into perspective: evolution, blind spots, and promising avenues. In: Donovan J, Stoian D, Hellin J, editors. Value Chain Development and the Poor: Promise, Delivery, and Opportunities for Impact at Scale. Rugby, UK: Practical Action Publishing; 2020

[4] Rutgers T. An analysis of supply side constraints on Ethiopian red pepper and paprika capsicum production and export: a global value chain approach 2010

[5] Haq ZU. Food value chain analysis: a review of selected studies for Pakistan and Guidelines for Further Research. Working Paper. 2012

[6] Hawkes C, Ruel M. Value chains for nutrition. 2020 Conference: Leveraging Agriculture for Improving Nutrition and Health. 2020 Conference Paper 4. 2011

[7] Kaplinsky R, Morris M. A Handbook for Value Chain Research. Institute of Development Studies, University of Sussex and School of Development Studies, University of Natal; 2001

[8] FAO. Developing Sustainable Food Value Chain. Guiding Principles. Rome, Italy: FAO; 2014

[9] Barnes M. Value Chain Guidebook: a Process for Value Chain Development. Canada: Agriculture and Food Council of Alberta Value Chain Initiative Alberta; 2004

[10] Gereffi G, Fernandez-Stark K. Global Value Chain Analysis: a Primer. North Carolina, USA: Center on Globalization, Governance & Competitiveness (CGGC), Duke University; 2011

[11] Webber CM, Labaste P. Building Competitiveness in Africa's Agriculture: a Guide to Value Chain Concepts and Applications. Washington DC: The World Bank; 2009

[12] IFPRI. Global Food Policy Report: Building Inclusive Food Systems. Washington, DC: International Food Policy Research Institute; 2020. DOI: 10.2499/9780896293670

[13] Elliot D, Gibson A, Hitchins R. Making markets work for the poor: rationale and practice. Enterprise Development and Microfinance. 2008;**19**(2):101-119

[14] Devaux A, Torero M, Donovan J, Horton D. Agricultural innovation and inclusive value-chain development: A review. Journal of Agribusiness in Developing and Emerging Economies. 2018;**8**(1):99-123. DOI: 10.1108/JADEE-06-2017-0065

[15] Mitchell J, Keane J, Coles C. Trading up: how a value chain approach can benefit the rural poor. London: COPLA Global: Overseas Development Institute; 2009

[16] Brewer BD. Global commodity chains & world income inequalities: The missing link of inequality and the upgrading paradox. Journal of World-Systems Research. 2011;**17**(2):308-327

[17] Neary JP. Putting the "new" into new trade theory: Paul Krugman's

Nobel Memorial Prize in economics. Scandinavian Journal of Economics. 2009;**111**(2):217-250

[18] Lin J, Chang HJ. Should industrial policy in developing countries conform to comparative advantage or defy it? A debate between Justin Lin and Ha-Joon Chang. Development Policy Review. 2009;**27**(5):483-502

[19] Faße A, Grote U, Winter E. Value chain analysis – methodologies on the context of environment and trade research: Diskussionspapiere der Wirtschaftswissenschaftlichen Fakultät der Universität Hannover No DP 429; 2009. Available from: www3.wiwi.uni-hannover.de/Forschung/Diskussionspapiere/dp-429.pdf [Accessed: 18 July 2012]

[20] van Melle C, Coulibaly O, Hell K. Agricultural value chain development in West Africa: Methodological framework and case study of mango in Benin. In: AAAE Conference Proceedings. 2008. pp. 49-52

[21] Porter ME. Competitive Advantage: Creating and Sustaining Superior Performance. New York: Free Press; 1985

[22] Roduner D. Analysis of existing theories, methodologies and discussions of value chain approaches within the development cooperation sector. A Draft Report. 2004

[23] Wallerstein I. The rise of the states-system: Sovereign nation-states, colonies, and the interstate system. In: World-Systems Analysis: An Introduction. New York, USA: Duke University Press; 2004

[24] Feller A, Shunk D, Callarman T. Value chains versus supply chains. BP Trends. 2006;**1**:1-7

[25] Mentzer JT, DeWitt W, Keebler JS, Min S, Nix NW, Smith CD, et al. Defining supply chain management. Journal of Business Logistics. 2001;**22**(2):1-25

[26] Burgess K, Singh PJ, Koroglu R. Supply chain management: a structured literature review and implications for future research. International Journal of Operations & Production Management. 2006

[27] Ketchen DJ Jr, Hult GTM. Bridging organization theory and supply chain management: the case of best value supply chains. Journal of Operations Management. 2007;**25**(2):573-580

[28] Halldorsson A, Kotzab H, Mikkola JH, Skjøtt-Larsen T. Complementary theories to supply chain management. Supply Chain Management: An International Journal; 2007

[29] Williamson OE. Transaction cost economics and the Carnegie connection. Journal of Economic Behavior & Organization. 1996;**31**(2):149-155

[30] David RJ, Han SK. A systematic assessment of the empirical support for transaction cost economics. Strategic Management Journal. 2004;**25**(1):39-58

[31] Williamson O, Masten S. The Economics of Transaction Costs. Cheltenham: Edward Elgar Publishing; 1999

[32] Stavins RN. Transaction costs and tradeable permits. Journal of Environmental Economics and Management. 1995;**29**(2):133-148

[33] Bachmann R, Witteloostuijn AV. Analysing inter-organizational relationships in the context of their business systems: A conceptual framework for comparative research. Reihe Soziologie. 2006;**78**:1-22

[34] Williamson OE. Comparative economic organization: The analysis

of discrete structural alternatives. Administrative Science Quarterly. 1991;**36**(2):269-296

[35] Dorward A, Poole N, Morrison J, Kydd J, Urey I. Markets, institutions and technology: missing links in livelihoods analysis. Development Policy Review. 2003;**21**(3):319-332

[36] Arega D, Manyong VM, Omanya G, Mignouna H, Bokanga M, Odhiambo GD. Smallholder marketed surplus and input use under transactions costs: Maize supply and fertilizer demand in Kenya. In: 2007 Second International Conference, Accra, Ghana. No. 52074. 2008

[37] Barrett CB. Smallholder market participation: concepts and evidence from eastern and southern Africa. Food Policy. 2008;**33**(4):299-317

[38] Ruben R, Boselie D, Lu H. Vegetables procurement by Asian supermarkets: A transaction cost approach. Supply Chain Management: An International Journal. 2007;**12**(1):60-68

[39] Chen H, Daugherty PJ, Landry TD. Supply chain process integration: a theoretical framework. Journal of Business Logistics. 2009;**30**(2):27-46

[40] Trienekens JH. Agricultural value chains in developing countries; a framework for analysis. International Food and Agribusiness Management Review. 2011;**14**(2):51-83

[41] Kilelu CW, Klerkx L, Leeuwis C. Supporting smallholder commercialisation by enhancing integrated coordination in agrifood value chains: experiences with dairy hubs in Kenya. Experimental Agriculture. 2017;**53**(2):269-287

[42] Berti G, Mulligan C. Competitiveness of small farms and innovative food supply chains: the role of food hubs in creating sustainable regional and local food systems. Sustainability. 2016;**8**(7):616

[43] Humphrey J, Schmitz H. How does insertion in global value chains affect upgrading in industrial clusters? Regional Studies. 2002;**36**(9):1017-1027

[44] Kim Y, Choi TY, Yan T, Dooley K. Structural investigation of supply networks: a social network analysis approach. Journal of Operations Management. 2011;**29**(3):194-211

[45] Coe NM, Dicken P, Hess M. Introduction: global production networks—debates and challenges. Journal of Economic Geography. 2008;**8**(3):267-269

[46] Gereffi G. Global commodity chains: New forms of coordination and control among nations and firms in international industries. Competition and Change. 1996;**1**(4):427-439

[47] Gereffi G, Lee J. Economic and social upgrading in global value chains and industrial clusters: why governance matters. Journal of Business Ethics. 2016;**133**(1):25-38

[48] Messner D. Regions in the 'world economic triangle'. In: Local Enterprises in the Global Economy. Cheltenham: Edward Elgar; 2004. pp. 20-52

[49] Kaplinsky R, Morris M, Readman J. The globalization of product markets and immiserizing growth: lessons from the South African furniture industry. World Development. 2002;**30**(7):1159-1177

[50] Gereffi G. Global value chains in a post-Washington Consensus world. Review of International Political Economy. 2014;**21**(1):9-37

[51] Humphrey J, Schmitz H. Governance and Upgrading: Linking Industrial Cluster and Global Value Chain Research. Brighton: Institute of Development Studies; 2000

[52] Sturgeon TJ. How globalization drives institutional diversity: the Japanese electronics industry's response to value chain modularity. Journal of East Asian Studies. 2007;**7**(1):1-34

[53] Ponte S, Sturgeon T. Explaining governance in global value chains: a modular theory-building effort. Review of International Political Economy. 2014;**21**(1):195-223

[54] Williamson OE. Transaction-cost economics: the governance of contractual relations. The Journal of Law and Economics. 1979;**22**(2):233-261

[55] Maertens A, Barrett CB. Measuring social networks' effects on agricultural technology adoption. American Journal of Agricultural Economics. 2013;**95**(2):353-359

[56] Lee J, Gereffi G, Beauvais J. Global value chains and agrifood standards: challenges and possibilities for smallholders in developing countries. Proceedings of the National Academy of Sciences. 2012;**109**(31):12326-12331

[57] Ganesh Kumar C, Murugaiyan P, Madanmohan G. Agri-food supply chain management: literature review. Intelligent Information Management. 2017;**9**:68-96

[58] Bardhan P, Mookherjee D, Tsumagari M, editors. Middlemen Margins and Globalization. Seminar Organized by IFPRI: Mimeo; 2009

[59] Abebe GK, Bijman J, Royer A. Are middlemen facilitators or barriers to improve smallholders' welfare in rural economies? Empirical evidence from Ethiopia. Journal of Rural Studies. 2016;**43**:203-213

[60] Giziew A, Negatu W, Wale E, Ayele G. Constraints of vegetables value chain in Ethiopia: a gender perspective. International Journal of Advanced Research in Management and Social Sciences. 2014;**3**(12):44-71

[61] Munshi K. Community networks and the process of development. Journal of Economic Perspectives. 2014;**28**(4):49-76

[62] Crona B, Nyström M, Folke C, Jiddawi N. Middlemen, a critical social-ecological link in coastal communities of Kenya and Zanzibar. Marine Policy. 2010;**34**(4):761-771

[63] Ponte S, Kelling I, Jespersen KS, Kruijssen F. The blue revolution in Asia: upgrading and governance in aquaculture value chains. World Development. 2014;**64**:52-64

[64] onte S, Ewert J. Which way is "up" in upgrading? Trajectories of change in the value chain for South African wine. World Development. 2009;**37**(10):1637-1650

[65] Humphrey J. Upgrading in global value chains. Available at SSRN 908214. 2004

[66] Douthwaite B, Hoffecker E. Towards a complexity-aware theory of change for participatory research programs working within agricultural innovation systems. Agricultural Systems. 2017;**155**:88-102

[67] Chamberlain. Smallholder Farmer Support Analysis Part 2: Theory of Change – Business model design for smallholder farmer support. Netherlands Enterprise Agency (RVO). 2019

[68] Rajala E, Vogel I, Sundin A, Kongmanila D, Nassuna-Musoke MG, Musundire R, et al. How can agricultural research translation projects targeting smallholder production systems be strengthened by using theory of change? Global Food Security. 2021;**28**:100475

[69] McKague K, Siddiquee M. Making Markets More Inclusive Lessons from

CARE and the Future of Sustainability in Agricultural Value Chain Development. Basingstoke, United Kingdom Palgrave MacMillan; 2014

[70] Bokelmann W, Adamseged ME. Contributing to a better understanding of the value chain framework in developing countries. In: 5th International Conference of the African Association of Agricultural Economists, Addis Ababa, Ethiopia. 2016

[71] Singh S. Making markets more inclusive: lessons from CARE and the future of sustainability in agricultural value chain development. Indian Journal of Agricultural Economics. 2016;**71**(1):90

[72] Bebbington A. Capitals and capabilities: a framework for analyzing peasant viability, rural livelihoods and poverty. World Development. 1999;**27**(12):2021-2044

[73] Alene AD, Manyong VM, Omanya G, Mignouna H, Bokanga M, Odhiambo G. Smallholder market participation under transactions costs: maize supply and fertilizer demand in Kenya. Food Policy. 2008;**33**(4):318-328

[74] Sen AK. From income inequality to economic inequality. Southern Economic Journal. 1997;**64**(2):384-401

[75] Solesbury W. Sustainable Livelihoods: a Case Study of the Evolution of DFID Policy. London: Overseas Development Institute; 2003

[76] Scoones I. Sustainable rural livelihoods: A framework for analysis. In: IDS Working Paper. Vol. 72. Brighton: IDS; 1998

Chapter 2

The Importance of Short Food Supply Chain: Examining Logistics Strategies for Competitive Advantage and Sustainability

Chris Dominic

Abstract

The Short Food Supply Chain (SFSC) has increased in recent decades, significantly impacting the environment, CO_2 emissions, and biodiversity loss. This study focuses on the SFSC and its components by examining the logistics strategies and the value of simple communication strategies in the SFSC. The study aims to gain insight into how these strategies can provide competitive advantages for food producers and align with the desired perception by customers. The SFSC is a direct and shortened food distribution system from producer to consumer, reducing the number of intermediaries involved in the supply chain and promoting a more sustainable and efficient food supply chain system. The study analyses case studies of Swedish food producers and intermediaries to understand the impact of SFSC on reducing food waste and packaging research. The growth of SFSCs is expected to lead to a more locally produced food industry and improved small business development, product range, and energy efficiency.

Keywords: short food supply chain, packaging, logistics, locally produced food, sustainability

1. Introduction

In recent decades, the Short Food Supply Chain (SFSC) has grown faster than the countries' GDP [1]. The SFSC significantly impacts the environment, CO_2 emissions, eutrophication, and biodiversity loss [2]. According to Eurostat 2009, road freight transport will increase by approximately 16% annually. The distance increase and the number of deliveries between the countries [3]. Meat consumption within Europe has also increased. About 365 million animals were transported within the E.U. market, of which about 67% were transported by road. The increased quantity and distance significantly impact the load on the environment, safety and quality. The transport sector generates approximately 25% of greenhouse gas emissions related to worldwide energy consumption. 75% of greenhouse gas emissions come from road transport [4, 5].

Increased transport, in turn, impacts the environment [6], logistics costs, quality and safety [7]. The demand for locally produced food has recently increased [8, 9]. The efforts to mitigate climate change risk [10] from transport related to the agricultural sector should receive more attention. The Swedish agricultural industry [11] uses about 20% of the total energy consumption. Effective utilisation of SFSC would reduce fossil fuels and greenhouse gases [12]. In addition, new national goals of self-sufficiency [13], creating logistics services, infrastructure solutions, freight transport systems and design challenges. Coordination of locally produced foods would improve competitiveness and quality for customers and consumers [14, 15].

This study focuses on Short Food Supply Chain and its components by examining the logistics strategies in the short food supply chain (SFSC), focusing on communication, environmental sustainability, and innovation. The study aims to gain insight into how these aspects can provide competitive advantages for food producers. Through case studies on a few Swedish food producers, intermediaries, and point-of-sale, the study analyses the value of simple communication strategies and how SFSC can align with the producers' desired perception by customers. The study does not consider the financial implications of SFSC. However, one key factor is to reduce food waste [16] and to service providers in logistics activities.

The ecological problems related to waste have led to [17] focus on packaging research that has increased [18–20]. On the other hand, packaging has a protective and informative role in showing the environmental impact [21, 22] for SFSC. However, several studies demonstrate the indirect importance of packaging that reduces environmental impact [23–25].

2. Short food supply chain

Short Food Supply Chain (SFSC) is a concept in food systems that refers to a direct and shortened food distribution system from producer to consumer. It emphasises local production and distribution, reducing the number of intermediaries involved in the supply chain [26]. SFSC aims to create a more sustainable and efficient food supply chain system by reducing the environmental impact of food transportation and promoting closer relationships between producers and consumers.

Short food supply chains (SFSCs), as legally defined by (E.U.) Regulation N. 1305/13 is considered a model of agricultural production that can achieve environmental, economic and social benefits, such as mitigating marginalisation inefficiencies and reducing transportation costs and CO_2 emissions.

SFSCs sales-end can be divided into individual direct sales, collective direct sales and locally organised associations. The main goal is to reduce the middleman and direct transactions between the farmer and the consumer. These transactions can occur inside or outside the farm, for example, at farmers' markets within a particular relation with the customer or in a collective form, involving cooperatives of producers selling their products to consumer groups, informal agreements and variable quantity levels between producers and consumers. Traditional SFSCs tend to be farm-based in rural areas and take the form of on-farm sales, roadside sales and 'pick-your-own' systems. Most typically, designated pages on social media are applied to inform the consumers about what is in the farmers' stock and how the items pack and distributed to rural marketplaces. SFSC management covers the entire chain of activities from production on the farm to processing, distribution, and marketplace to the consumer.

Locally produced food means that the food is grown or produced close to where you make your purchases. What is considered close is not defined, but in Sweden, close is usually regarded as regional or local. Choosing locally produced food benefits producers and businesses in the home area and reduces the environmental impact that long transports would otherwise entail. But it should be mentioned that the concept does not necessarily positively impact the environment.

2.1 SFSC local associations

Several non-profit member associations have emerged supporting local farmers with the mission to increase the volume and quality of food production in various parts of Sweden. The association's goal is for those who live in the region to partake of the locally produced food. They have mapped all food producers within the country so that one, as a private person, can know where to shop, experience and eat food produced in local areas.

The association aims to respond to the growing interest and demand for locally produced food requiring increased production and processing. This development must also be ecologically, economically and socially sustainable in the long run. According to one member, the growth of SFSCs will lead to more employment. The ripple effect will improve the development of small businesses, increase a broader product range, increase attractiveness for the rural area, and a more energy-efficient society will positively affect the hospitality industry, expand research activities and a higher level of education.'

2.2 Sales channels

New sales channels with short supply chains have grown as a reaction to long-distance food supply chains. The discussion about SFSCs raises the question of finding alternative ways to supply food by reducing the distance between restaurants and private customers or consumers. In recent years the use of digital ordering systems for buying food has increased in combination with the increased interest in consuming locally produced food.

2.3 Virtual food platform

In Sweden, the virtual food platform "REKO-Ring" has been popping up, connecting consumers and producers interested in selling and buying locally-produced food using social media groups.

Reko-Ring is a sales channel that food producers can use to sell their products. With a predetermined date and place, all transactions are done in about an hour, where possible payment and pick-up of goods occur.

The Reko-Ring can be said to be a mobile farm shop where all business is settled in advance, which reduces the amount of food waste and saves the producers a lot of time as the period for picking up the products is predetermined. Moreover, producers can plan deliveries in a time-efficient and environmentally friendly way and coordinate these instead of going to several locations. The orders are made through the local Reco-Ring social media group, where the producers have their payment approach.

2.4 Farm shop

Farm shop means that the producer has his shop, usually on his farm, where the producer's goods, often self-made, are sold as other belongings. The purchase then occurs locally, with the transaction and the goods being given directly to the hand at the time of sale.

2.5 Community kitchen

The community kitchen means food produced for publicly owned kitchens serving preschools, elderly care, healthcare, and hospitals. This is another interesting sales channel for locally grown food, as public opinion exists for this type of set-up system. The local authority tenders locally produced food and distribution activities to benefit small-scale businesses.

2.6 Retail

Grocery stores mean Swedish retail, such as, e.g. ICA, Coop, Willys and Hemköp. Also, these actors are working closely to benefit the local food industry.

2.7 Public market halls

A market hall is a public sales outlet where products such as meat, fish and seafood, fruit and vegetables, delicacies, spices and flowers are sold.

2.8 Packaging

Packaging influences all logistics operations throughout SFSC and helps the customer in decision-making. Well-functioning packaging systems play an increasingly vital component in succeeding in creating a well-functioning logistics process. The packaging design is crucial to transporting the goods from local farms to customers in a thoughtful, smooth, and cost-effective way. With well-designed packaging, waste and transport are reduced, which in turn means that costs for producers are reduced.

Standardisation is needed to achieve a functioning logistics chain with simple packaging for handling and use. It is vital to jointly develop a maximum packaging size and weight for smooth handling for the producer, transport company, and customers.

3. Case study

3.1 Logistics model for locally produced food

One study was conducted in the western region of Sweden. The study had various hubs/central warehouses where producers deliver their products themselves or a third-party logistics company, such as transport and last-mile distribution companies. The goods from several local food producers were collected with a separate shipping company. The thought process here was to reduce the environmental impact and cost and make it possible for the producers to use the shipping company.

From the various hubs, the goods were transported to the customers with the help of the shipping company, and these deliveries were made directly to the customer's desired location.

Locally produced is a resource centre for small-scale food producers that advise, train and develops small-scale food companies to create growth and employment in the region. The vision is to become the leading region in locally produced food. It currently has seven hubs that deliver to customers and consumers. In addition, three different transport companies were used depending on the goods to be sent.

3.2 Reko-rings in the central region of Sweden

A site visit was made when Reko-Rings had an event where the producers and customers met for sales and picked up groceries. Approximately five local food producers came there, and all five used their means of transport. One producer cooperated with another and pooled farmed products. The means of transport used by all producers were fossil-fueled cars, but there was an interest in a more environmentally friendly alternative.

One of the food producers took payment on the spot. While with the remaining four producers, the goods had already been prepaid for in advance by the customers. One of the producers had surplus food with him and was, therefore, able to sell beyond his predetermined orders.

An observation that was made was that there was a low amount of customers concerning what the producers wanted. Several people pointed this out during the interviews.

3.3 Small-scale fishery

In Grundsund, on the west coast of Sweden run, fishermen have a small-scale coastal fishery. In this area, small-scale fishing held the key to living in the archipelago.

The longitudinal canals leading into the well-protected harbour are houses and boathouses. The fishing gear on the wooden quays testifies to the professional fisherman. Here is a fisherman who believes in a bright future in the profession, and fishing becomes a lifestyle.

- The fisherman says we still have a living archipelago with small-scale fishing.

He used to work in trawling but has been fishing with cages primarily for lobster and crayfish for several years. Occasionally, he also engages in net fishing for flounder, mackerel, and other types of fish.

He has his mother, a small-scale professional fisherman, with him out to sea-his father also fishes, which he has done since the 1980s. During parts of the fishing season, he brings tourists from a nearby hotel. The tourist then learns more about what type of species swim in these parts and life as fishermen.

He grew up with small-scale coastal fishing, but for a few years, he tried a slightly different type of fishing in Scandinavia.

- It is simply this fishing that is closest to my heart. But, I also have the opportunity to come home daily to my family, different from those out fishing from Sunday to Thursday or Thursday to Tuesday, which is a regular work schedule.

He takes the herring fishery in the southern part of the Baltic sea as an example of how large-scale fishing can wipe out small-scale fishing.

Once upon a time, herring fishing generated work for many families in southern parts of Sweden. Nowadays, most of the herring is caught by a large trawler, as they catch in Denmark.

- Even if it is fishermen from these parts with Swedish-owned boats flagged, it generates either more work for more people on land in Sweden or tax revenue for Sweden.

- From the authorities' point of view, it may be more manageable to control 10 larger boats than 100 smaller ones.

The interviewee still sees a clear swing that speaks for small-scale fishing. More private individuals, restaurant buyers, and fish markets are asking for fish caught in cages. Traceability becomes an essential factor when customers are demanding cage fishing.

3.3.1 Protected zones for sustainable fishing

The fisherman advocates protected zones where cage fishing can occur during parts of the year. This fishing method has little impact on the bottom's environment and animal and plant life. It also offers more people an opportunity to make a living from small-scale fishing that is sustainable in the longer term.

Another advantage is that shellfish that do not measure up are thrown back alive. But, unfortunately, that opportunity is not given to the animals caught in single trawls, where bycatch is thrown back dead into the sea.

Because large-scale fishing fleets are often prioritised in fisheries policy centres, it is easier for the E.U. to control and fit them like a glove in many situations. But large fishing boats trawling far out at sea do not contribute similarly to the old, classic fishing modes.

- There are no more fishing boats out there. And the prawns come frozen from Greenland. So then we understand that fishing and vessels are essential in this, states the local fisherman.

The interviewee has a positive vision of the future of this sector. The possibility of changing the direction of fishing and aiming it more precisely at selected species also reduces the pressure on fish species that need to be protected. In this way, the stocks can grow larger and more robust.

- In some regions, combining overburden fishing and trawl fishing may not be possible. But for us, as it is here now, I think it works well. Of course, from a sustainable point of view, trawl fishing should cease, limit, or move out elsewhere. However, we are optimistic that many will still be able to adapt.

4. Discussion

Short Food Supply Chain (SFSC) refers to a direct and shortened food distribution system from local producer to consumer, reducing the number of intermediaries involved. It aims to create a more sustainable and efficient food supply chain system by reducing the environmental impact of food transportation and promoting closer relationships between producers and consumers. SFSCs can be individual direct sales, collective direct sales, or locally organised associations. New sales channels for SFSCs have emerged, such as virtual food platforms, i.e. Reko-Ring and farm shops, retail stores, and public market halls. Packaging plays an essential role in SFSC, influencing all logistics operations and helping customers in decision-making.

This study presents two case studies on short food supply chains and small-scale fishing conducted in central and western parts of Sweden. The case studies were conducted on the spot by mapping the short food supply chain channel from the producers to the point of sale. In the first case, local food producers meet for sales and deliver at an event, with five participating suppliers in the study. As a result, the producers face low customer turnout and interest in a more environmentally friendly means of transport. In the second case, a fisherman and his family run a small-scale fishery, focusing on cage fishing for lobster and crayfish. The fisherman advocates for protected zones for sustainable fishing and has a positive outlook on the sector's future, as there is a growing demand for traceable fish caught in cages.

The study highlighted the challenges of large-scale fishing and its impact on the environment and the overfishing of certain species.

The Short Food Supply Chain (SFSC) is a direct and shortened food distribution system from producer to consumer that reduces the environmental impact of food transportation and promotes closer relationships between producers and consumers. The trend has been to minimise distribution and transport distances by bringing more locally consolidated food production and becoming more self-sufficient. The interest in locally produced food led to the development of non-profit member associations supporting local farmers and new sales channels like the virtual food platform. The increased use of SFSC can lead to improved employment, small business growth, reduced food waste, and lower greenhouse gas emissions. In addition, SFSC can offer a more sustainable and efficient food supply system by decreasing the distance between food producers and consumers.

5. Conclusions

The study focuses on Short Food Supply Chain (SFSC) and its impact on logistics strategies, communication, environmental sustainability, packaging, and innovation. The study aims to understand the competitive advantages of SFSC for food producers through case studies of Swedish food producers, intermediaries, and point-of-sale. The SFSC exemplifies a direct and shortened food distribution system that emphasises local production and distribution and reduces the number of intermediaries involved. It aims to create a more sustainable and efficient food supply chain system by reducing the environmental impact of food transportation and promoting closer relationships between producers and consumers. There are three types of SFSCs: individual direct sales, collective direct sales, and locally organised associations. The growth of SFSCs could lead to more employment opportunities, improved small businesses, and a low-emission food supply for consumers.

Author details

Chris Dominic
Industrial Management and Sustainability, Gävle, Sweden

*Address all correspondence to: chris.dominic@hig.se

References

[1] Curtis F. Eco-localism and sustainability. Ecological Economics. 2003;**46**:83-102

[2] EEA. Environmental indicator report 2016—In support to the monitoring of the 7th environment action programme. European Environmental Agency. 2018;**1**:1-34. Available from: https://www.eea.europa.eu//publications/environmental-indicator-report-2018. EEA Report No 19/2018. [Accessed November 2022]

[3] Crippa M, Solazza E, Guizzardi D, Monforti-Ferrario Tubiello FN, Leip A. Food systems are responsible for a third of global anthropogenic GHG emissions. Nature Food. 2021;**2**:198-209. DOI: 10.1038/s43016-021-00225-9

[4] Intergovernmental Panel on Climate Change (IPCC). Climate change and land - An IPCC special report on climate change, desertification, land degradation, sustainable land management, food security, and green gas fluxes in terrestrial ecosystems. Summary for Policymakers Approved Draft. 2019;**1**:1-874. Available from: https://www.ipcc.ch/site/assets/uploads/2019/08/Fullreport-1.pdf. [accessed November 2019]

[5] Määttä-Juntunen H, Antikainen H, Kotavaara O, Rusanen J. Using GIS tools to estimate CO_2 emissions related to the accessibility of large retail stores in the Oulu region, Finland. Journal of Transport Geography. 2010;**19**(2):346-354

[6] Marletto G, Sillig C. Environmental impact of Italian canned tomato logistics: National vs. regional supply chains. Journal of Transport Geography. 2014;**34**:131e141. DOI: 10.1016/j.jtrangeo.2013.12.002

[7] Rothwell A, Ridoutt B, Page G, Bellotti W. Environmental performance of local food: Trade-offs and implications for climate resilience in a developed city. Journal of Cleaner Production. 2016;**114**:420-430. DOI: 10.1016/j.jclepro.2015.04.096

[8] Engelseth P. Developing exchange in short local foods supply chains Miewald. International Journal on Food System Dynamics (Pembroke, Ont.). 2016;**7**(3):1-14. DOI: 10.18461/ijfsd.v7i3.734

[9] Ugaglia AA, Del'homme B, Lemarié-Boutry M, Zahm F. Reflets et Perspectives de la Vie Economique. 2020. DOI: 10.3917/rpve.591.0019

[10] Godde CM, Mason-D'Croz D, Mayberry DE, Thornton PK, Herrero M. Impacts of climate change on the livestock food supply chain; a review of the evidence. Global Food Security. 2021;**28**:1-17. DOI: 10.1016/j.gfs.2020.100488

[11] Sivertsson O, Tell JS. Barriers to business model innovation in Swedish agriculture. Sustainability (Switzerland). 2015;**7**:1957-1969. DOI: 10.3390/su7021957

[12] Nordmark I, Ljungberg D, Gebresenbet G, Bosona T, Jüriado R. Integrated logistics network for the supply chain of locally produced food, part II: Assessment of E-trade, economic benefit, and environmental impact. Journal of Service Science. 2012;**5**:249-262. DOI: 10.4236/jssm.2012.53030

[13] Dachner N, Tarasuk V. Canadian Food Studies/La Revue canadienne des études sur l'alimentation. 2018. DOI: 10.15353/cfs-rcea.v5i3.278

[14] Markar'yants SB. The development of agriculture in Japan and the issue of food security. Ekonomika Sel'skokhozyaistvennykh i Pererabatyvayushchikh Predpriyatii. 2010;**5**:82-86

[15] Zarei M, Fakhrzad MB, Paghaleh JM. Food supply chain leanness using a developed QFD model. Journal of Food Engineering. 2011;**102**:25-33

[16] Elnakib S. A National strategy to reduce food waste at the consumer level. Journal of Nutrition Education and Behavior. 2021;**53**(7):551-640. DOI: 10.1016/j.jneb.2021.04.003

[17] Chen S, Brahma S, Mackay J, Cao C, Aliakbarian B. The role of smart packaging system in food supply chain. Journal of Food Science. 2020;**85**(3):511-853. DOI: 10.1111/1750-3841.15046

[18] Beitzen-Heineke Elisa F, Nazmiye B-O, Reefke H. The prospects of zero-packaging grocery stores to improve the social and environmental impacts of the food supply chain. Journal of Cleaner Production. 2017;**140**(3):1528-1541. DOI: 10.1016/j.jclepro.2016.09.227

[19] Tencati A, Pogutz S, Moda B, Brambilla M, Cacia C. Prevention policies addressing packaging and packaging waste: Some emerging trends. Waste Management. 2016;**56**:35-45

[20] Matar C, Salou T, Hélias A, Pénicaud C, Gaucel S, Gontard N, et al. Benefit of modified atmosphere packaging on the overall environmental impact of packed strawberries. Postharvest Biology and Technology. 2021;**177**:1-9. DOI: 10.1016/j.postharvbio.2021.111521

[21] Molina-Besch K, Wikström F, Williams H. The environmental impact of packaging in food supply chains—Does life cycle assessment of food provide the full picture? International Journal of Life Cycle Assessment. 2019;**24**:37-50. DOI: 10.1007/s11367-018-1500-6

[22] Majewski E, Komerska A, Kwiatkowski J, Malak-Rawlikowska A, Was A, Sulewski P, et al. Are short food supply chains more environmentally sustainable than long chains? A life cycle assessment (LCA) of the eco-efficiency of food chains in selected E.U. countries. Energies. 2020;**13**:1-26. DOI: 10.3390/en13184853

[23] Battini D, Calzavara M, Persona A, Sgarbossa F. Sustainable packaging development for fresh food supply chains. Packaging Technology and Science. 2016;**29**(1):25-43. DOI: 10.1002/pts.2185

[24] Silvenius F, Grönman K, Katajajuuri JM, Soukka R, Koivupuro HK, Virtanen Y. The role of household food waste in comparing environmental impacts of packaging alternatives. Packaging Technology and Science. 2014;**27**:277-292

[25] Wikström F, Williams H, Venkatesh G. The influence of packaging attributes on recycling and food waste behaviour—An environmental comparison of two packaging alternatives. Journal of Cleaner Production. 2016;**137**:895-902. DOI: 10.1016/j.jclepro.2016.07.097

[26] Paciarotti C, Torregiani F. Short food supply chain between micro/small farms and restaurants: An exploratory study in the Marche region. British Food Journal. 2018;**120**(8):1-13. DOI: 10.1108/BFJ-04-2018-0253

Chapter 3

Trends Shaping the Future of Agrifood

John Stanton and Rosa Caiazza

Abstract

In agriculture innovation is the main driver of productivity growth. Innovative technologies and methods have to help increase firms' productivity in a sustainable way. Technology diffusion is important as developing new technologies for sustainable production processes. Current innovations can lead to define the composition of foods with impoverishment or enrichment of nutritional and sensory characteristics. New technologies play a central role in adapting agricultural practices to environmental change. New agricultural practices help to preserve environmental integrity. Competition for an alternative use of natural resources is increasing and agricultural practices and technologies will have to adapt to climate change and more extreme weather-related conditions. This multifunctional approach needs to be directed towards increasing knowledge about the relationship between microstructure, process, product characteristics and technological innovation to protect the quality and origin of products. Innovations are not enough if they are not accompanied by a regulatory environment conducive to business development. Think first of all about the issues of security, rationalization of controls and the embargo of bureaucracy. But also, to the sustainability of food production and the need for common rules on supply conditions in crucial international commodity markets for the development of certain products.

Keywords: supply chain, agribusiness research, agribusiness trends, agribusiness pillars, distribution

1. Introduction

Agriculture refers to the production of food and non-food items through farming or animal husbandry [1]. It encompasses the realization of both agricultural products (the rearing of livestock and the growing of crops and horticulture), as well as agricultural services (agricultural animal husbandry and horticultural services and the operation of irrigation systems). It is highly dependent on natural resources (land, soil, climatic conditions and water) that differ significantly across the countries, with implications for the pattern of industry, investment and trade [2]. It also requires private investmentments in innovations and institutional support for technological improvements. According to FAO world agricultural production will grow (from 10 to 20%) in the next years if policy-makers will realize institutional and technological support to firms involved in agriculture.

Policy-makers play a crucial role in agricultural extension, development of new seed varieties and supporting agricultural production. The government can contribute to such support by providing agriculture-related infrastructure facilities aimed to increase the productive capacities of firms. However, the extent to which institutions contribute to agricultural production varies by country and by type of institution. Budgetary constraints in developing countries limit their capacity to establish relevant institutions in support of agricultural development.

Technological improvements play an important role in increasing agricultural productivity. Modern agricultural technologies can help increase agricultural productivity in a more sustainable way. In agriculture, innovation is the main driver of productivity growth [3]. It can also improve the environmental performance of firms and the quality of products. Some innovations help firms better deal with production uncertainties and increase profits. In some countries, the challenge is to adapt agricultural production to consumer needs.

Consumers' behaviors are changing according to living standards, higher participation rates of women in the labour force, and reduced time available for meals [4]. Innovation must address such changes in consumers' behaviors, provide raw materials for non-food use, alleviate natural resource depletion, and enable adaptation to the expected changes in natural conditions caused by climate change [5]. Developed countries invest considerably more in agricultural innovations than developing countries. Therefore, it is essential to increase public budgets for both institutional and technological support to competitiveness of firms involved in all the stages of the global value chain in the food industry [6].

2. General framework

The global value chain in agribusiness comprises the suppliers of inputs, processors of agricultural goods and retailers. It assumes a global form involving both local or foreign players. In general, value creation resides mainly in the non-agricultural segments of agribusiness chains. Some global value chains are coordinated by transnational corporations (TNCs). The universe of agriculture-related TNCs, including suppliers of inputs, food manufacturers [7] and retailers [8] are usually larger than agricultural TNCs [9]. The United States is home to the largest number of food processing TNCs (ex. Kraft Foods and Coca-Cola), followed by the United Kingdom, Netherlands, and Switzerland [10]. In the developing world, Hong Kong (China), Singapore and Mexico are the most important home economies of food processors. Behind the TNCs, the food processing industry is populated by a huge number of small and medium-sized firms (SMEs) that must compete through specialization in high-quality products and affiliation to a global network [11].

To face the challenges of an increasingly global market, small and medium-sized firms need to base innovations on the local tradition to meet the changing needs and lifestyles of consumers. Thanks to such investments SMEs that realize traditional products (preserves, cheeses, wine, oil, etc.) are more and more involved in the production of new goods (frozen, ready-made sauces, fresh condiments, etc.). Most of SMEs associate technological innovation (new products integrated with new production processes) to non-technological innovations (organization or marketing). The growth strategy of many SMEs has to focus on strengthening the markets in which they already operate and entering new markets through organizational innovation in networks with other actors. In implementing both technological and non-technological innovations in the agro-food industry, firms have to consider the new trends of sustainability, healthy and smartness proposed in the '2030 Agenda' from the United Nations. The 2030 Agenda for Sustainable Development identifies Sustainable Development Goals, with a view to stimulating action over the next 15 years in all economic, social and environmental areas of critical importance for sustainable development (**Table 1**) [12].

As stated in the 2030 Agenda, an important characteristic of the agro-food industry is its close association with the environment. Food production contributes to climate change, water scarcity and the destruction of biodiversity. Agro-food firms have to avoid environmental degradation through pollution, greenhouse gas emissions and deforestation.

The first challenge facing the food industry is to intensify production by reducing the environmental impact and pressure on natural resources, but at the same time producing safe products that allow people to have a diet that ensures good health. The objective of firms has to promote a development model based on the balance between environmental, economic and social needs. The sustainability of agri-food production contributes to economic, social and environmental development. Firms have to extend the reuse of materials, bioenergy, more efficient use of energy and water in the production processes of foods. Policy-makers must stimulate the development of models for sustainable production that make agro-food firms able to satisfy the present and future needs of the world's population [13].

The second challenge facing the food industry is to reduce health problems. In developed countries, an increasing number of people have health problems due to what is commonly referred to as lifestyle (ex. mainly diet, physical activity, mental and emotional stress, etc.). People's lifestyle has had huge changes in the last decades. Social and economic change due to globalization has had a strong impact on society leading to pervasive changes in people eating habits and behaviors (ex. eating out of the home, personalized diets, consumer preferences, e-commerce, etc.).

Another challenge is to reduce toxic substances that residue from ingredients used in production processes or move from the environment to the food chain [14]. Great attention should also be paid to the so-called emerging contaminants (organostannic biocides, brominated flame retardants, perfluorinated organic compounds) reported by scientific literature for food but not yet considered by law. To address this second challenge, it is necessary to use multidisciplinary strategies able to combine advanced technologies to offer safe products with consumers' information [15]. Food security is now linked to innovation and sustainability of production systems and is perhaps the most important element of business competitiveness [16]. The research must cover all phases of the value chain from the production to the final consumption (from farm to fork) [17].

The last challenge is to become smarter and smarter in the organization of firms' activities. Collecting and transforming more and more data into strategic and usable information is not a process sufficient to ensure greater rationality in economic

Value chain	Suppliers	Processors	Retails
Innovation	Technological		Non-technological
New trends	Sustainability	Health	Smartness

Table 1.
Pillars of agro-food system.

decisions [18]. The third challenge is to improve firms' ability to know how to fully exploit the generated information. From this point of view, the agri-food sector is ideally suited to take full advantage of information and communication technologies (ICT) in all the stages of the value chain for both specific activities (ex. supply, processes, marketing, distribution, sales, retail, etc.) and support activities (ex. technology transfer). ICT in the agri-food industry aims to enable firms to become more and more efficient in realizing managerial purposes using data available throughout the value chain. This allows for a natural metamorphosis of business models, production processes and redefinition of relationships between the members of the chain. With the aim to evidence main innovations to face such challenges an empirical analysis of innovative brands was realized.

3. Empirical analysis

The empirical analysis evidenced that innovation is not limited to just one category or just one company. The chart indicates that innovation is spread across both. Many of these innovations are related to the challenges (**Table 2**).

The packaging we use is the best way to keep our crisps crunchy. However, the new consumers looking for new products want to see more sustainable packaging, 100% recyclable packaging and still be delicious. TerraCycle is trying to make this a reality.

Philadelphia Cream Cheese has focused on the third challenge by solving the problem of technology in that they can now make 13 different flavors of cream cheese such as Salmon, Pumpkin spice, honey nut, chive, onion, etc. Bird's eye also demonstrated a focus on Health by introducing steaming pouching that permits cook-in-the-bag technology. New Covent Garden bringing chilled soup to market for the first time a technology-driven innovation. Genius Foods makes Gluten-free bread where the focus is not on just the gluten but that it is tastier, softer and smells delicious. This type of innovation captures both perceived health and consumer attributes not generally met by gluten foods. Nontechnical innovation can be seen in the perception of brands as fun and vibrant and appears more likely to be influenced

Brand	Main market the brand operates in	Innovative brands %
Walkers	Crisps	46
Müller	Yogurt	45
Ben & Jerry's	Ice cream	42
New Covent Garden	Soup	41
Birds Eye	Frozen food	39
Philadelphia	Cream cheese	38
Haribo	Sugar confectionery	38
Genius	Gluten-free baked goods	38
Magnum	Ice cream	37
Activia	Yogurt	37

**Base: Approximately 2000 internet users aged 16+.*

Table 2.
Innovative brands.

+10%	+16%	+23%	+17%	+22%	+15%
Cereals	Meat	Dairy products	Fish	Sugar	Ethanol

Table 3.
Agricultural production outlook.

by the category they operate in, rather than inherent perceptions of being innovative. Brands like Müller, Ben & Jerry's, Magnum and Walkers all operate in treat categories, which lends an image of fun and vibrancy to the brands compared to those operating in more staid and functional markets. The new social media has prompted more interest in new products. One in 10 are prompted to buy new products by social media and 12% of food shoppers are encouraged to buy a new food product by related social media content from brands. However, this is low in comparison to the usage of social media. A total of 67% of food buyers are prompted to buy a new food when it has been recommended by a friend or family member and 47% of adults like to recommend new food products to others. This indicates that any steps to drive word of mouth could be used to support new product launches. It is ironic that the new social media and technology have made the old "word of mouth" more effective [19]. Value chain at all levels is one of the most important factors in the successful introduction of a new product about this factor all three levels are shown in **Table 3**. The number one factor in encouraging a trial purchase is a price promotion. In this case, every level from Suppliers, processors and retailers all have the obligation to keep costs down and be prepared to use marketing dollars to support promotional prices [20]. The processors can invest in technology that makes the crops more efficient to produce, processors can not only rely on technology but use trade promotion dollars to reduce the prices at retail and finally retailers must be willing to pass on the promotional money that was designated to reduce the retail price and therefore the risk [21].

As previously mentioned, social media can play a huge role in supporting new products. Recommendations from family and friends are the third most important factor to encourage trial. A recommendation by a friend or family member would prompt two-thirds of shoppers to buy a new food product. This underlines the value of word of mouth to brands when launching new products [22]. There is no question that this media is difficult as it is new, and many older companies are not comfortable with this media. Additionally, remains challenging, since it is difficult to measure the impact. The online snack delivery brand Graze, for example, has offered discounts to customers when their friends/family sign up [23]. Other brands have taken advantage of digital platforms to encourage recommendations. For example, Birds Eye set up a pop-up restaurant in London in the spring of 2014, which let people pay by taking a picture of their meal and posting it to Instagram. Similarly, when Special K launched its Cracker Crisps, it set up a pop-up shop that allowed people to pay for the crisps with tweets.

4. Results

From the research, we defined a general framework that considers the trend of a sector, time, complexity, and degree of uncertainty as the variability underlying the forecasting choices. The forecast time depends on the static or dynamic nature of the sector being analyzed. Mature industries with rather stable dynamics tend to be slow to change and therefore allow forecasts over longer time frames. On the contrary, the sectors most exposed to innovations have faster dynamics of change which allow forecasts over shorter time frames. The degree of complexity depends on the number of variables to consider. The greater the number of relevant variables, the greater the complexity of forecasting the trend of the sector. Finally, the degree of uncertainty depends on the level of knowledge of the relevant variables. The greater the knowledge of the relevant variables, the lower the degree of uncertainty in forecasting the trend of the sector (**Figure 1**).

To apply the model identified to the study of future trends in the tomato sector, a medium-long time period is considered, taking into account the maturity of the agri-food sector, to limit complexity, innovation, sustainability and internationalization and to reduce the degree of uncertainty, each of them is operationalized into two macro-categories. This model must take into consideration a series of main agri-food trends such as technological and non-technological innovation, the environment and sustainability, the individual and corporate choices for internationalization and entry into new markets (**Figure 2**).

In the production and processing industry, we must focus on the joint activity of technological and non-technological innovation. Technological innovation capable of determining substantial improvements in the quality of the product or of the production process must increasingly be accompanied by innovation non-technological relating to organization and marketing activities. The competitiveness of the sector must, in fact, be based on the constant research and implementation of products to respond to new needs, the introduction of new processes that consider the growing digitization of the economy and new organizational forms and marketing actions capable of significantly improving the national or international adoption of the manufactured products [19].

The primary objective of the agri-food processing industry is to offer good, healthy, and safe food at affordable prices for everyone. To this end, companies in the sector must be engaged in product and process innovations aimed at creating healthier products suitable for different needs and lifestyles. The know-how starts from an adequate selection of the best raw materials available and is carried out through production processes born from ancient gastronomic traditions [20]. The agricultural and agri-food sector are in the ideal conditions to fully exploit the potential offered by information and communication technology to individual companies and to the production chains in which they are inserted, both as regards specific activities (such as research and development), both as regards support activities (such as technology transfer). In parallel with product or process innovations, companies also implement marketing innovations aimed at helping the consumer make choices based on

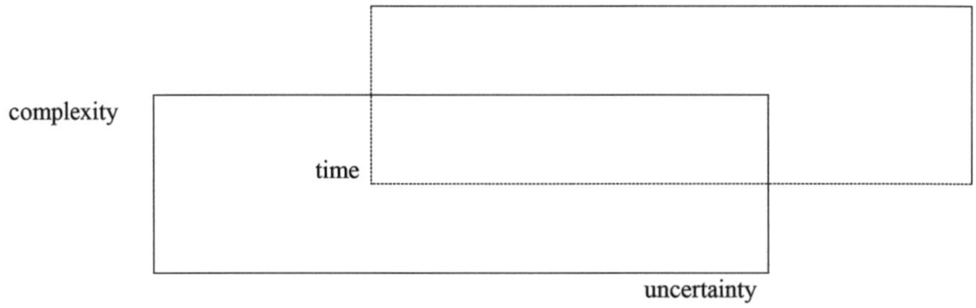

Figure 1.
Model.

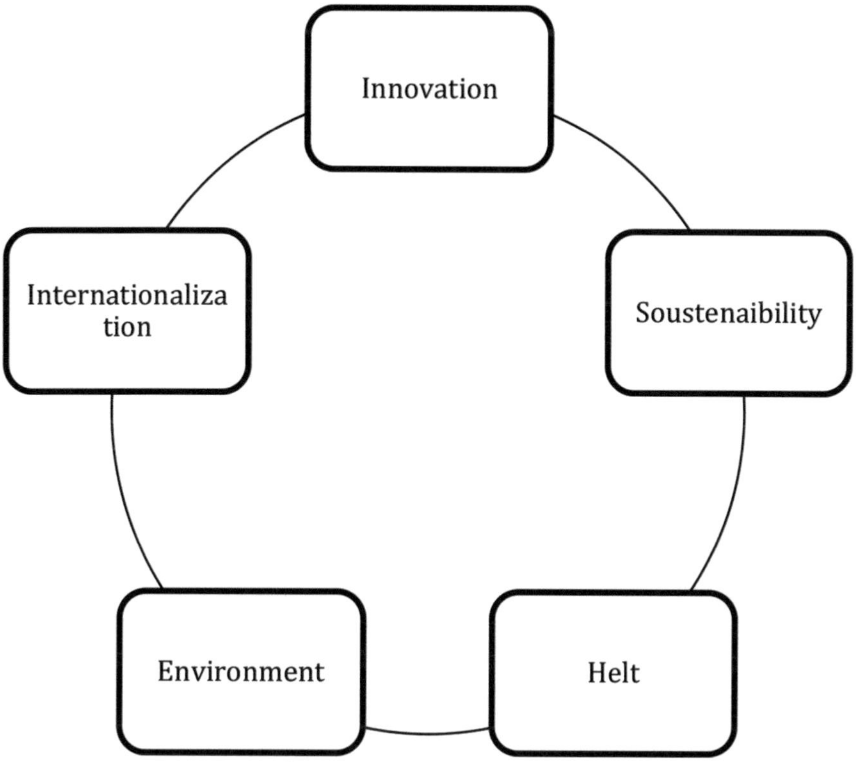

Figure 2.
New trends.

detailed information on products, raw materials and conservation methods. These innovations must above all concern packaging which plays a key role in guaranteeing food quality and safety, protecting product integrity during transport, distribution and consumption, conveying brand values and also providing nutritional and service information essential for the consumer [24]. The work of consumer information and communication must be accompanied by real innovation on the market to guarantee new solutions suitable for different lifestyles and targets. The improvement of the nutritional characteristics of food products is achieved through a modification of their composition, as far as technologically possible and accepted by the consumer and by trying to maintain the organoleptic characteristics of the product (flavor, texture and shelf life). Most agri-food companies that intend to compete in the global market choose the development of innovations in design and the adoption of new solutions in the field of wrapping and packaging that enhance the new products as a strategy for diversifying and improving the production offer [25]. The growth strategy of many companies is based on organizational innovations which provide for the strengthening of the distribution network in the markets in which they already operate or the entry into new markets by setting up their own sales networks abroad [26]. Further innovations concern the organization of the network of raw material suppliers. In order to be able to create products to be introduced on the global market, the Italian food industry is not self-sufficient for the raw materials of some product chains. In these types of matters, it is necessary to implement innovations of an organizational nature aimed at optimizing relations with foreign suppliers through integrated management systems which provide for participation in the activities upstream of the supply chain.

Companies in the agri-food sector should direct their activities towards the development on a global scale of production and consumption business models capable of making Italian companies capable of meeting the present and future needs of the world population while respecting the environment and communities territorial. From this point of view, the objective of companies extends from offering quality products to doing so with respect for the environment and future generations. The sustainable development of agri-food companies must consider the possibility of extending as much as possible the policies for the reuse of waste by-products, in the production of bioenergy, in the cosmetic and pharmaceutical industry and in the production of fertilizers. Significant progress must also be made in terms of more efficient use of energy and water in production processes along the entire food chain.

The issue of sustainability must be accompanied by that of health and food safety. The lifestyle of people worldwide has undergone profound changes in a relatively short period of time. In fact, in the last decade, social and economic changes due to globalization have had a strong impact on the organization and functioning of our society, leading to pervasive and consistent changes in eating habits and behaviors. In the meantime, a real demographic revolution has taken place which has led to a substantial increase in life expectancy. This change in eating habits and lifestyles has resulted in a greater orientation of the consumer towards food products with health benefits and of better quality. This has led to a natural segmentation of the market which must lead companies to consider the needs of new market niches that respond to the needs of the elderly, vegans, sportsmen, etc.

Environmental issues related to the agricultural sector are influencing the way farmers and businesses in the sector operate in agricultural production through the adoption of more sustainable and environmentally friendly cultivation techniques, better management of water resources, the use of renewable energy sources. The extensive use of chemicals and pesticides in agriculture has polluted rivers, lakes and other water resources and adversely affected the health of agricultural workers. The conversion of forests to agricultural land has a significant impact on biodiversity and the destruction of wildlife and its habitats.

The international growth strategies of agri-food companies can be aimed at defending existing markets or at entering new markets and can be implemented in different ways depending on the objectives that the company sets itself to achieve. From this point of view, it is necessary to distinguish the determinants underlying the internationalization choices from the methods of implementation of these strategies. In the first case, it is necessary to identify which factors drive towards the choice of a given market, which opportunities are present in it or which reasons make it impossible to exploit the same opportunities in the domestic market. In the second case, however, it is necessary to focus on the decision-making processes of companies, on the reasons for choosing one method rather than another and on the main resources that make a given internationalization path possible. Traditionally, internationalization processes in the agri-food sector have involved large companies. In recent decades, however, many small and medium-sized enterprises (SMEs) have carried out their economic activities outside national borders, becoming international players. SMEs face more challenges than large multinationals to compete successfully globally. The main barriers to the internationalization of SMEs are, in fact, the scarcity of working capital to finance exports, the difficulty in identifying foreign business opportunities, the limited information to analyse the markets, the inability to contact with potential clients, the difficulty in obtaining certain foreign representation, the lack of time of managers who mainly deal with internationalization, the inadequate amount of personnel not trained in internationalization. These problems can be traced to a general lack of time and managerial skills to internationalize, financial means and knowledge of foreign markets. The success of an internationalization operation requires the prior definition of a clear strategy and the acquisition of a series of skills, abilities and resources. The greater number of constraints and the substantial differences in the available resources and organizational structures of SMEs compared to large multinationals highlight a series of limitations in using the literature of the latter to explain the choices of the former.

5. Conclusion

Relationships between microstructure, process, product characteristics and technological innovation can protect the quality and origin of products. Innovations are not enough if they are not accompanied by a regulatory environment conducive to business development. Policy-makers could play a leading role in deploying this innovative approach. In developing countries, the challenge is to use mostly internal resources to develop inexpensive technologies and enrich food formulations. In developed countries, the need is to balance scientific innovation with cultural changes. Harmonization with the regulatory framework and homogeneous implementation of land-based regulations are key concepts to ensure the proper development of efficient agro-food policies. Think first of all about the issues of security, rationalization of controls and the embargo of bureaucracy. But also, to the sustainability of food production and the need for common rules on supply conditions in crucial international commodity markets for the development of certain products.

Author details

John Stanton[1] and Rosa Caiazza[2]*

1 Saint Joseph's University, USA

2 Parthenope University of Naples, Italy

*Address all correspondence to: rosa.caiazza@uniparthenope.it

References

[1] Audretsch DB, Caiazza R. Technology transfer and entrepreneurship: Cross-national analysis. The Journal of Technology Transfer. 2015;**41**(6):1247-1259

[2] Capitanio F, Coppola A, Pascucci S. Product and process innovation in the Italian food industry. Agribusiness: An International Journal. 2010;**26**:503-e518

[3] Van Trijp HCM, Meulenberg MTG. Marketing and consumer behaviour with respect to foods. In: Meiselman HL, McFie HJH, editors. Food Choice, Acceptance and Consumption. London: Blackie Academic and Professional; 1996

[4] Caiazza R. Benchmarking of business incubators. Benchmarking: An International Journal. 2014;**21**(6):1062-1069

[5] West J, Gallagher S. Challenges of open innovation: The paradox of firm investment in open-source software. R&D Management. 2006;**36**(3):319-e331

[6] Aguilera JM. Seligman lecture 2005: Food product engineering: Building the right structures. Journal of the Science of Food and Agriculture. 2006;**86**:1147-e1155

[7] Caiazza R, Audretsch DB, Volpe T, Singer J. Policy and institutions facilitating entrepreneurial spin-offs: USA, Asia and Europe. Journal of Entrepreneurship and Public Policy. 2014;**3**(2):186-196

[8] Caiazza R, Richardson A, Audretsch DB. Knowledge effects on competitiveness: From firms to regional advantage. The Journal of Technology Transfer. 2015;**40**(6):899-909

[9] Caiazza R. Explaining innovation in mature industries: Evidence from Italian SMEs. Technology Analysis & Strategic Management. 2015;**27**(8):975-985

[10] Caiazza R. A cross-national analysis of policies effecting innovation diffusion. The Journal of Technology Transfer. 2015;**41**(6):1406-1419

[11] Bercovitz JEL, Feldman MP. Fishing upstream: Firm innovation strategy and university research alliances. Research Policy. 2007;**36**(7):930-e948

[12] Bonney L, Clark R, Collins R, Fearne A. From serendipity to sustainable competitive advantage: Insights from Houston's farm and their journey of co-innovation. Supply Chain Management: An International Journal. 2007;**12**(6):395-e399

[13] Tyler JEI. Advancing university innovation: More must be expected more must be done. Minnesota Journal of Law, Science & Technology. 2009;**10**:143-e212

[14] Traitler H, Watzke HJ, Saguy IS. Reinventing R&D in an open innovation ecosystem. Journal of Food Science. 2011;**76**(2):R62–eR68

[15] Candelin-Palmqvist H, Sandberg B, Mylly UM. Intellectual property rights in innovation management research: A review. Technovation. 2012;**32**:502-e512

[16] Sarkar S, Costa AIA. Dynamics of open innovation in the food industry. Trends in Food Science and Technology. 2008;**19**:574-e580

[17] Saguy IS. Paradigm shifts in academia and the food industry required to meet innovation challenges. Trends in Food Science & Technology. 2011;**22**:467-e475

[18] Caiazza R, Stanton JL, Volpe T. Global supply chain: Consolidators'

role. Operations Research Perspectives. 2016;**3**:1-4

[19] OECD. Global Food Security: Challenges for the Food and Agricultural System. Paris: OECD Publishing; 2013. DOI: 10.1787/9789264195363-en

[20] Huston L, Sakkab N. Connect and develop: Inside Procter and Gamble's new model for innovation. Harvard Business Review. 2006;**84**:58-e66

[21] Costa AIA, Dekker M, Beumer RR, Rombouts FM, Jongen WMF. A consumer-oriented classification system for home meal replacements. Food Quality and Preference. 2001;**12**:229-e242

[22] Omta SWF, Folstar P. Integration of innovation in the corporate strategy of Agri-food companies. In: Jongen WMH, Meulenberg MTG, editors. Innovation in Agri-Food Systems. Wageningen: Wageningen Academic Publishers; 2005. pp. 223-e246

[23] OECD and FAO (Food and Agriculture Organization). OECD-FAO Agricultural Outlook 2016-25. Paris: OECD Publishing; 2016. DOI: 10.1787/agr_outlook-2016-en

[24] Mikkelsen BE, Kristensen NH, Nielsen T. Innovation processes in large scale public foodservice e case findings from implementation of organic foods in a Danish county. Journal of Foodservice Business Research. 2005;**8**:87-e105

[25] Knudsen MP. The relative importance of interfirm relationships and knowledge transfer for new product development success. Journal of Product Innovation Management. 2007;**24**:117-e138

[26] Menrad K. Innovation in the food industry in Germany. Research Policy. 2004;**33**(6e7):845-e878

Chapter 4

Technology Promotion and Scaling in Support of Commodity Value Chain Development in Africa

Paul L. Woomer, Rachel M. Zozo, Sabra Lewis and Dries Roobroeck

Abstract

Strengthening the production and processing of key food commodities forms the basis of agricultural development in Africa. These value chains follow a quasi-linear progression across seven main segments: farm planning > land preparation and crop establishment > field production > harvest > post-harvest handling > marketing > and value addition. Each of these consists of sub-segments whose improvement depends upon promotion and adoption of specific modernizing technologies. The technologies either have commercial application, as with the distribution of production input products and labor-saving equipment, or are related to management of farms and processing. For crop commodities, these products include improved varieties planted with more and better-formulated fertilizers and pest management materials. Management options are primarily directed toward the better conservation of resources and wiser integration of different farm enterprises. Key factors underlying value chain advancement include wider application of digital services, more effective incentives for climate-smart action, increased mechanization and irrigation, improved marketing efficiency and fairness, and incentives for value-creating agro-processing. An analogous set of factors also relate to value chains supporting animal enterprise. Attracting women and youth to meaningful careers in agriculture is particularly important since they are major stakeholders in the scaling of much-needed technologies and business models.

Keywords: cassava, farm mechanization, goats, sheep, technology transformation, wheat

1. Introduction

From a business point of view, value chains provide a means to describe how inputs and services are combined to design, grow, and transform a product. They also describes how products move from producers to consumers; and how they increase in value along the way. Originally, the value chain concept was used to understand business relationships [1], with the main purpose of achieving and maintaining competitive advantage. Later, definitions of value chains were expanded to the complete range of activities for advancing a product through its phases of production, including linkages between similar stage producers, and links to other value chains that provide needed goods and services [2]. The term was quickly adopted within agricultural development circles to guide their interventions [3]. In this context, value chains serve as a reference point for efforts designed to reduce poverty and a means to direct benefits toward them through a variety of initiatives focused on improved productivity and the growth of stakeholder enterprises. The concept grew into the common wisdom that value chains work best when their actors cooperate to generate ever-increasing benefits for all participants working in concert to create value and build trust. This view contrasts greatly from producers operating in more competitive and adversarial modes. At this point, value chains had grown from a useful way to understand manufacture and sales into a solution toward positive social and economic change. The chains were viewed more in terms of the actors operating within them rather than the processes they conduct. In the case of agricultural development, this entails actors with interests in social equity and economic justice in addition to various suppliers, farmers, traders, processors, distributors, and consumers.

Hill [4] proposed three transitional stages toward more sustainable agricultural systems: (i) efficiency; (ii) substitution; and (iii) redesign. The 'Efficiency Stage' focuses on more efficient allocation of existing farm resources and judicious use of available production inputs. It recognizes that many agricultural systems are inherently wasteful, and much is gained from reducing losses and increase efficiency. The 'Substitution Stage' involves the strategic replacement of technologies and practices. It allows for the introduction of new crop varieties and livestock breeds and encourages suites of accompanying technologies that greatly increase productivity. The 'Redesign Stage' focuses on agro-ecosystem function to achieve sustainable impacts at scale [5]. It advocates for recent and future breakthroughs that not only increase productivity, but also manage externalities in positive ways such as greenhouse gas emissions, carbon sequestration, soil and water quality, biodiversity, and dispersal of pests and diseases [6]. Note that 'Efficiency' and 'Substitution' are readily applied to existing production systems, whereas 'Redesign' involves future transformative changes within them. For agricultural value chains to have relevance within the scope of progressive change and improvement, they must consider elements of all three transitional stages. To a large degree, this can be achieved through incremental substitution of more efficient technologies promoted and adopted by various value chain actors. Within the context of development, each transitional stage is best supported through government investment as addressed in Sections 8 and 9 of this paper.

In this Chapter, we examine value chains in terms of the technologies that are available to their actors. This approach is specifically applied to ongoing efforts toward the transformation of African agriculture that places modernization at the very center of economic growth [7, 8]. Under these programs, a host of production and processing technologies is widely disseminated to smallholder farmers and entrepreneurs across major agro-ecological zones [9]. Interventions are based on proven technologies for key food commodities, their positional linkages within value chains, and the accrual of benefits by wider acceptance and practice of such technologies. Most notably, in this way a bridge is formed between the more conventional approach to value chains as consortium of actors [10] and the staged, modernizing transition toward agricultural improvement [4].

2. Technology-based value chains

A generic value chain model for supporting modernized crop-based agriculture appears in **Figure 1**. This chain consists of seven main segments: farm planning, crop establishment, seasonal production, harvest, post-harvest, marketing, and value addition. Consumers are not depicted but they operate at the extreme right of this sequence. Each segment contains three links, corresponding to different technical requirements. Note that planning allows for long-term, seasonal, and immediate access to input and labor allocations. Crop establishment consists of three stages; land preparation, soil fertilization, and planting, although additional early operations may be required. Crop management considers three basic operations: weeding, supplemental fertilization, and pest and disease control. Crop harvest involves the timing, mode of recovery and means of removal from the field. Primary post-harvest operations occur immediately after the crop is collected and involve the separation of edible parts from residues, bagging and storage of produce, and various quality preservation measures intended to protect commodity value. Marketing involves contact with and delivery to buyers, product inspection to ensure expected quality, and the financial transactions through which goods are purchased. In some cases, these buyers or their clients may be processors where secondary products are manufactured, packaged, and distributed. It is important to note that each of these segments are supported by agricultural and food processing technologies and that these technologies improve production and processing efficacy, allowing for substitution of inputs and management strategies over time. Long-term planning allows ecosystem refinements as described by Hill [4] as well as iterative substitution by more advantageous and affordable inputs.

Along these key agricultural commodity value chains, critical "enablers" advance agricultural transformation. Among these are greater use of digital tools for timely and accurate decision-making; judicious implementation of tax benefits and other government incentives for wider scale agro-ecological goals such as those related to climate action; investments in irrigation and mechanized practice for reducing risks and drudgery; greater protection from harvest losses; and improved access to markets, value added processing and exporters. Again, each of the modernization objectives is supported by a "technology toolkit" that allows for substitution of input products and management practices that reward land managers and practitioners over time.

Overall, the value chain for animal enterprises is very similar to that of crop enterprises (**Figure 2**) but has important differences in its details. For instance,

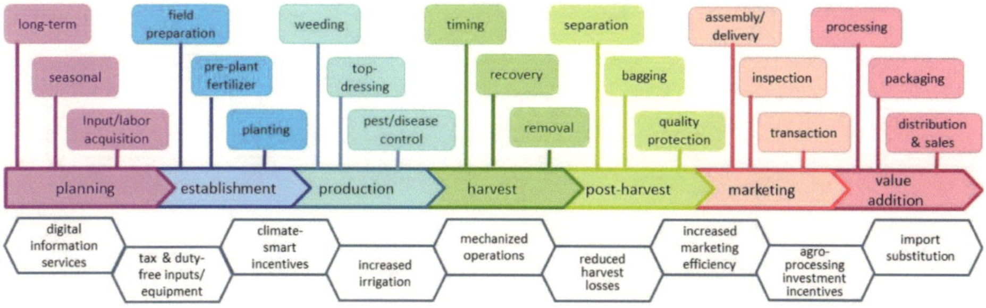

Figure 1.
Modernized generic agricultural crop value chain and expected outcomes.

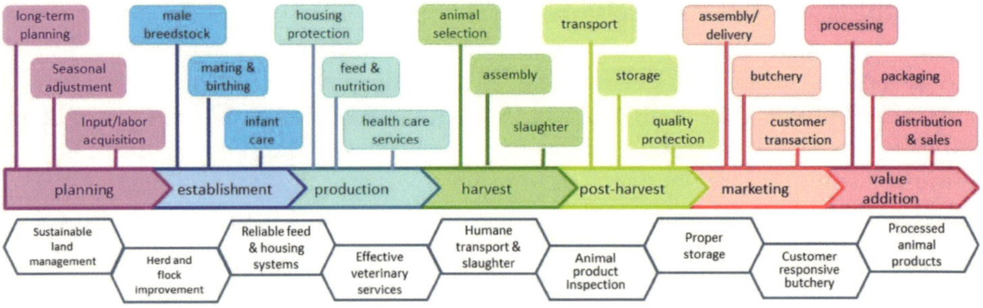

Figure 2.
Modernized generic animal enterprise value chain and expected outcomes.

long-term planning relates to the stewardship of large rangeland areas under increasing threats of degradation by climate change. Similar concerns exist for waste disposal and greenhouse gas emissions from poultry houses and fishponds. Breeding strategies are also enacted over many years. Seasonal adjustments include the needs for containment structures and improved feeding systems. Production inputs such as vaccines and feed supplements require short-term and recurrent consideration. Unlike for crops, animal establishment, production and harvest are not necessarily sequential, most particularly for birds and fish. Herd improvement is continuous in nature but mating and birthing of livestock are seasonal. Demand for housing protection, feed and health care remain constant throughout the year. Animal slaughter involves more stringent moral and public health considerations than for crop value chains. Because livestock produce is readily perishable, post-harvest, marketing and value addition are closely linked to cold storage. The distinctive implications of modernizing technologies for small-ruminant enterprises in Africa are explained later in this Chapter.

While this paper primarily examines value chains from the perspective of the technologies they deliver and how they change over time, it does not ignore the key actors within them such as farmers and their workers, traders and wholesalers, and processors and retailers. These primary actors are owners of raw, semi-processed or finished products, steadily moving them toward consumers. Equally important are actors that assist in agricultural planning and investment, producer associations and the manufacture, distribution and sales of production inputs and equipment. As a rule, the more complex a value chain and its associated technologies, the more actors it engages. When producers sell directly to consumers, its value chains are simpler and actors fewer, becoming increasingly longer and more complex when goods are

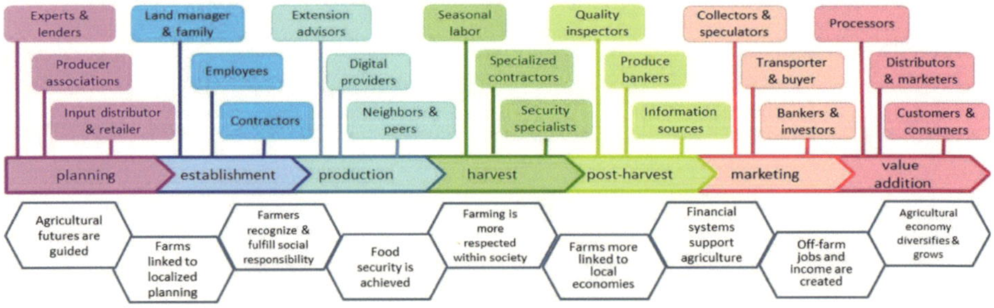

Figure 3.
Key value chain actors within a developmental context and their various roles.

transported, traded, and processed across greater distances. Farmers must understand the roles of other actors within their value chains, so they work effectively with them. In the context of this Chapter, this particularly relates to private sector actors who distribute and market modernizing technologies. Generally, actors most able to benefit from value chains are more entrepreneurial and better communicators. **Figure 3** describes the important value chain actors and their various roles. Upstream actors have the responsibility of securing agricultural futures and advancing new production technologies while downstream actors ensure product quality, add value, and anticipate consumer needs.

3. Cassava technologies

Cassava is one of the most important crops for food and nutritional security in the humid African tropics. It is also an agro-industrial commodity for processing and trade within world markets. Cassava farmers in Sub-Saharan Africa suffer yield losses of 12–23 million ton of fresh roots per year worth US $1200 to $2300 million [11]. The main reasons for shortfalls in production are disease and pest damage, reliance upon unimproved planting material, inappropriate fertilizer management and poor weed control. Further value is lost due to the untapped potentials for biofortification, improved post-harvest handling and agro-industrial processing. Ten technologies are recognized that serve to modernize production and processing of cassava in Africa [12].

3.1 Cassava varieties resistant to disease and pests

Production of cassava by farmers in Sub-Saharan Africa is widely limited by pernicious viruses such as cassava mosaic disease and cassava brown streak disease [13]. Many recently released improved varieties also withstand bacterial blight, green mite, and mealybug, offering major advantages to producers. For example, widespread adoption of improved varieties in southwestern Nigeria improves tuber yield by about 40% [14].

3.2 Higher dry matter and starch contents

The amount of dry matter and starch in cassava roots greatly influences their value in terms of agro-processing options. Cassava varieties are now available that have root dry matter contents of 40–45% [15]. Greater starch content translates into better nutritional value and higher selling price.

3.3 Golden-fleshed cassava

Conventional and advanced breeding approaches resulted in the increase of pro-vitamin A content in cassava, offering substantially improved nutrition within rural communities [16]. Roots of "golden" yellow-fleshed cassava are rich in a beta-carotenoid that give its characteristic color. The sales price for golden cassava roots on markets in Sub-Saharan Africa is up to 20% higher than that of white, non-biofortified types.

3.4 Stem bulking enterprise

Cuttings from cassava stems are the most commonly used planting material by African farmers because this kind of propagule can be gathered from previous crops, it sprouts quickly and reliably, and allows ever-increasing areas of land to be cultivated. Under optimal crop and soil management, it is possible to multiply cassava cuttings in only six to 10 months' time. In general, each cutting may be produced and marketed for as little as US $0.02 each, and sold for up to US $0.04 each, and up to 24 cuttings may be produced on 1 m^2, resulting in revenues of US $4800 per ha [12], above and beyond tuber yield.

3.5 Semi autotrophic hydroponic multiplication (SAH)

This technology involves turning cuttings of cassava roots into small plantlets. A tray of ready-to-market product typically contains 12 to 48 plantlets that can be covered, stacked, and packaged for transport. Multiplication facilities for SAH with an area of 40 square meters can produce 75,000 plantlets per month that are sold for US $0.07 to $0.10 per piece [12]. SAH allows for more rapid introduction of improved varieties.

3.6 Specially blended fertilizers

Mixes of common inorganic fertilizers have been specifically developed for cassava and other root crops that create balanced availability of nutrients for the crop's below ground production. Fertilizing cassava with the correct balance of nutrients at the right time and placement improves the productivity and quality of tubers, and strengthen resilience to drought, and pests and diseases, while avoiding undesired losses to the environment. Smaller, more labor-intensive blending systems may be developed for localized operations, and even operated as a community-based operation once specific formulations are known and component ingredients mobilized [12].

3.7 Mechanized planting and harvesting

Mechanical planters can plant 7 to 10 hectares daily, making it much faster and less expensive than manual planting. Similarly, mechanical lifters can harvest up to 3 to 5 ha of cassava in a day. Mechanical harvesters range in complexity from simple lifters to multipurpose tractor attachments that lift, shake and gather roots from multiple rows. Mechanical planting and harvesting greatly reduce labor bottlenecks that undermine cassava production [12].

3.8 Cassava weed management

The wide spacing and slow initial canopy development in stands of cassava make it susceptible to weed encroachment during early cultivation. The "Six Steps" approach is a complete package for weed management that incorporates multiple key control measures, including site selection, weed identification, herbicide application, tillage operations, plant spacing, and post-emergence weeding. Weed removal following this method costs US $28–46 per hectare. Well-weeded cassava provides 30–50% greater root yield than a poorly weeded operation [17].

3.9 Cassava peels for animal feed

Processing cassava roots into food products results in massive amounts of peels. Typically, 1 ton of fresh cassava roots results in 200–300 kg of peels. Peels made into mash or meal can serve as an ingredient for feed formulations of poultry, cattle, and other livestock. The crude protein content of cassava peel animal feeds is low, amounting to only 4–6% for wet meal, but the starch content is high at 77–78% [18]. Processing cassava peels provides a low-cost, energy-rich feed that is ideal for substituting higher-cost cereals in feed blends.

3.10 High quality cassava flour (HQCF)

Roots of cassava are made into HQCF through a series of industrial steps and is suitable for manufacturing a wide range of foods. Flour blending allows for partial substitution of wheat flour and other imported foods. HQCF requires fresh roots with high dry matter and starch content and needs to take place within a day or two after harvest. HQCF is a gluten-free product that can be used within non-allergenic foods, a market segment that attracts rapidly growing global demand [19].

4. Wheat technologies

Sub-Saharan Africa produces only 30% of its own wheat. This massive reliance upon imports creates a crippling but largely unnecessary trade imbalance, resulting in allocation of foreign reserves that might otherwise be directed toward more strategic developmental purpose. Geographic information, simulation modeling and economic analyses indicate that less than 10% of the potential for profitable wheat production is currently exploited on the continent [20]. Intensifying and growing this sector along the lines of agricultural redesign [4] requires changes in attitudes and policy, and developing reliable value chains for seeds, production inputs and equipment, and output markets. Ten technologies hold great promise for increasing the production and processing of wheat in Africa [21].

4.1 Heat and drought tolerance

Heat and drought stress are two major constraints to wheat production in Sub-Saharan Africa. Diurnal temperature spikes above 36°C undermine flowering and grain filling leading to low yields or crop failure. Recently developed varieties of wheat that withstand heat offer real opportunity to expand wheat production [22]. Farmers can now grow the crop in non-traditional areas such as drylands in the Sahel characterized by temperatures of 30–40°C and rainfall below 250 mm, or in savanna areas currently targeting maize. Costs for seed multiplication are low since the crop is self-pollinating.

4.2 Rust resistance

Yellow rust and stem rust are devastating diseases in major wheat production zones of Sub-Saharan Africa [23, 24]. Wheat varieties that are resistant to yellow rust and stem rust now exist that can be grown at the same planting density and agronomic recommendations as other cultivars. In some cases, fungicide spraying may also be required to ensure season-long protection of the crop. Producing wheat resistant to rusts costs about US $440 per hectare and grain yields commonly exceed 4 ton ha^{-1}.

4.3 Hessian fly control

Wheat in Northern Africa is threatened by infestations of the Hessian fly (*Mayetiola destructor*), and this pest is increasingly common in Sub-Saharan Africa as well [25]. The larva of this insect causes substantial losses, mostly by feeding on the growth apex. Planting wheat varieties that possess a natural defense mechanism against the Hessian fly larvae and releasing and promoting natural enemies are two highly effective control methods for this pest [26].

4.4 Expanded irrigation

Growing wheat during the cool season in African drylands, referred to as "winter" production, avoids adverse effects of heat stress and allows production of two or more crops in a year. Low rainfall during the cool season requires irrigation for which efficient options are available [27]. The success of irrigated wheat production in several "breadbasket" regions of Africa offers great promise to advance self-sufficiency and reduce expensive importation.

4.5 Raised bed cultivation

Furrow irrigated, raised bed cultivation is a technique that enhances water use efficiency and avoids waterlogging. Beds and furrows are relatively easy to construct with locally available tools and can be maintained for several growing seasons. Compared to conventional flood irrigation, water use efficiency in raised beds improves by about 25% as the wetted area is less [28]. These engineered surfaces also promote the collection and infiltration of rainwater and decrease soil erosion.

4.6 Conservation agriculture (CA)

This sustainable land use technique involves a set of soil and crop management practices that offer major advantages for wheat production [21]. The strategy has a low implementation cost, saves on fertilizer, labor, and irrigation, and provides higher, more stable yields and profits under reduced rainfall. Practicing CA is based on reduction of soil disturbance, residue retention of the previous crop, and timely weed and fertilizer management. Different types of no-till seeders are available, including manually or animal power devices and small to large tractor-drawn attachments. Adopting CA enriches soil biodiversity, reduces gaseous emissions, and sequesters carbon in soils, benefiting the environment and mitigating climate change.

4.7 Integrated pest management strategies (IPM)

Combinations of varietal, cultural, biological, and chemical methods for crop protection reduce yield damage and costs for farmers. IPM strategies involve carefully selected techniques that are tailored to local conditions. A wide range of biological measures are available that target pests. A full IPM package that includes increased fertilizer, precision herbicide application, and seed treatment costs about US \$515 ha^{-1} [29]. Adopting these strategically bundled technologies boosts average productivity to by 3.5–4.8 ton ha^{-1}.

4.8 Combine harvesters

Motorized equipment that cuts wheat crops and separates grain are available in a broad range of sizes, from small units that can handle a few hectares per day to very large units for major operations that harvest several hectares per hour. Smaller units with a cutting width of 1.2 m cost as little as US $12,000. Harvesting cost is about US $25 to $33 per hectare, or US $4.7–9.2 per ton of grain [30]. Costs of cutting, threshing, straw disposal, winnowing and overall grain losses during harvest are substantially reduced by combine harvesting.

4.9 Hermetic storage bags

This technology greatly reduces post-harvest loss of grain by using sealed bags that exclude air and moisture, creating an environment that is non-conducive for insects and rots. The bags preserve the quality of grains and obstruct the entry of insects and microorganisms through depletion of oxygen levels and accumulation of carbon dioxide. Threshed grains are dried and placed into high-density polyethylene bags. Wheat grain can be stored for up to 2 years with this preservation technology. The raw material and labor to manufacture hermetic grain storage bags costs between US $1 to $1.5 per piece and are sold at US $2 to $3 by suppliers [21].

4.10 Flour milling

Small- to large-sized equipment for milling and blending are available that allow manufacturing of premium wheat flour near to production areas. There is a wide range of equipment for sorting wheat based on size, shape, and density of grains, as well as cleaning and annealing. A mill with an output capacity of 300–500 kg flour per hour costs around US $3500. Fully automatic flour mills with a capacity of 30 ton flour per day are sold from US $38,000. Modern equipment achieves high recovery of flour at 80–82% and 18–20% bran, much higher than traditional techniques [21].

5. Small ruminant Enterprise

Raising goats and sheep is an important source of food and income across Africa. These types of livestock are easy to rear, multi-purpose and valuable, especially within subsistence and pastoral communities. Small ruminants supply protein in form of meat and milk, generate income for their owners and create employment for millions of others in the value chain as herders, traders, butchers and processors. Ten technologies are most important to intensify small ruminant production systems in Africa [31].

5.1 Herd improvement through community-based breeding

Traditional breeds of goats and sheep exhibit useful adaptation to environmental stress and partial resistance to common diseases but are often lower in meat and milk production compared with improved breeds. A community-based approach to breed improvement builds upon these traits of interest relying upon hardy stock performance [32]. Herds of selected ewes (female sheep) and does (female goats), and a few recognizably superior rams (male sheep) or bucks (male goats) are steadily improved through community-based action. This approach creates an enabling environment for goat and sheep enterprise that strengthens local cooperatives and market linkages.

5.2 Reduced overgrazing and rangeland rehabilitation

Rising population and increasing demand for animal products place excess pressure on land leading to overgrazing and rangeland degradation. Climate change exacerbates this situation, yet technologies exist to conserve and restore rangelands. Successful management involves monitoring the vegetative growth of grazing areas and level of water points; and regulating herd numbers and movement. Available technologies to combat rangeland degradation include rotational grazing, afforestation, fodder production, soil and water conservation, and policy interventions [33]. Appropriate stocking density reduces overgrazing and allows for land recovery.

5.3 Small ruminant containment

Housing protects animals from inclement weather and provides better feed, waste management and biosecurity conditions. A shed can be built from locally available materials such as timber, bamboo, or lumber off-cuts in combination with wire and fencing. These sheds usually include access to nearby daytime grazing. Proper shelter includes feed and water troughs that may be wooden, metal or plastic, sometimes fed through automated devices. Where farmers have access to inexpensive wooden posts and planks, a suitable shed can be constructed for as little as US $200. More elaborate buildings are constructed for about US $20 per m^2. Every young goat or sheep protected through improved housing reflects a savings of about US $150 [31].

5.4 Pasture improvement

Pastures are tracts of land producing fodder grass and other vegetation for confined and free-grazing animals. Usually such parcels are fenced and receive inputs such as fertilizers and irrigation. Several approaches are followed in the establishment of pastures such as control of weedy patches, partial land disturbance and sowing of improved grasses and legumes, under sowing croplands with grazing plant species allowing for the establishment of crop-pasture rotations, and establishment of shrub hedgerows along pasture margins. The cost of new pasture establishment is approximately US $500 per ha, offering additional feed worth between US $250 to $500 per ha per year over the next many years [31].

5.5 Cut-and-carry fodder systems

Cut-and-carry describes a system where feed is gathered and offered to confined animals. It facilitates more efficient feed management by reducing wastage; but also places greater demand on labor and nearby vegetation resources. It also secures maximum advantage from crop residues and seasonally available vegetation. Under this system, a lamb worth US $80 consumes fresh chop worth $30, supplements and medicines costing $40 and then produces meat worth $250 over six months [31].

5.6 Short-term fattening

Goat and sheep finishing involves intensive and nutritious feed regimes that promote fast growth, attaining desired carcass growth and quality. It maximizes the value of livestock with minimum time and space, which offers a business opportunity through value addition of purchased stock. It requires moderate investment and offers minimal risks; allowing peri-urban dwellers to become engaged in small ruminant value chains. Young adult animals are purchased and finished to slaughter weight by limiting their movement and providing them with a concentrated diet [34]. Fattening is readily visible, resulting in a profitable cycle of 3–6 months, often targeting festive seasons. This practice in turn creates increased demand for fattened animals and more rapid turnover of stock [34]. This operation may be repeated three times per year over several years once facilities are secured, offering an estimated profit margin of US $ 149 per sheep [31]. Fattening operations are associated with a transition toward raising larger animals [35] and increasing asset ownership [36].

5.7 Universal vaccination against diseases

"Peste des Petites Ruminants" (PPR) is a serious disease of goats and sheep across Africa. This is a fast-spreading viral disease with high mortality rates, especially among younger animals. Its symptoms are rapidly elevated body temperature, with affected animals displaying discharges from the eyes and nose, sores in the mouth, troubled breathing, coughing, and foul-smelling diarrhea [37]. There is no cure for PPR but vaccination is available to protect herds. The vaccine is thermostable and inexpensive, costing US $0.5 to $1 per animal [31]. It is also important that all producers be aware of this disease and isolate any infected animals at an early stage of development.

5.8 Manure processing

Goat and sheep manure has economic value but realizing market opportunities requires expertise. Some advantages of manure from small ruminants are its relatively high and balanced nutrient content, and naturally pelleted form. These may be applied fresh to soil without damaging plants, or as a mulch or compost ingredient. Commercial technologies are available to produce organic fertilizers from manure. After composting, production involves crushing, granulating, drying and screening for pellet uniformity. Processed manure is sold for US $200 to $1500 per ton depending on the level of processing and quality of packaging [31].

5.9 Humane slaughter

Humane slaughter refers to the killing of an animal instantly or rendering it insensible until death follows, without pain, suffering or distress [38]. When slaughtering animals for food, they must be stunned by electricity prior to bleeding out so they become quickly unconscious. Another approved method of stunning involves percussion bolt pistols. Humane slaughter is based upon principles of animal rights whereas those with utility as human food have moral worth and must be protected from unnecessary suffering. Modest investment of US $2000 is adequate to set up humane slaughter, allowing an abattoir operator to earn a profit of US $15 to $20 per animal as they operate in regulatory compliance [31].

5.10 Hide curing and secondary leatherworks

Hides are a valuable product from livestock production. For this value to be realized, animals and their hides must be properly treated, and artisans require skill sets and appropriate materials. Hides are processed by local communities, stockpiled, sold, and fabricated by leatherwork industries into a variety of products for both domestic and export markets. Foremost among those products are shoes, handbags, and leather clothing, with premium value obtained through greater craftsmanship. A modest investment of about US $1000 establishes a local leatherwork business [31].

6. Small-scale farm mechanization

Small-scale mechanization is the key to reducing drudgery and increasing productivity among African farmers. For too long, farming in Africa remains regarded as a path to poverty rather than a profitable agribusiness. This is due in large part for dependency upon tedious hand labor, poor returns to effort and lack of investment into farming systems. A wide range of small-scale equipment is available to improve returns to labor and change this poor image of farming. This equipment serves the entire value chain including tillers, power weeders, power sprayers, soil augers, irrigation systems, multi-crop threshers and a host of other equipment that are becoming commercially available for the first time [39]. An example of modernized post-harvest and processing technologies is the axial-flow thresher and GEM parboiling system. The thresher may be locally fabricated and results in high quality milled rice while the parboiler uses rice husk as fuel, is smokeless and produces a rapidly prepared grain [40]. Description of eight other important machinery follows.

6.1 Hand tractors for land preparation

Hand tractors are two-wheeled machines that range in power from 5 to 20 horsepower. They are most often attached to a rotavator that tills soil, but other attachments are available, including those that create furrows and plant and bury rows. These tractors are guided by handlebars that provide control over their direction. High-power hand tractors can break new ground and dry, crusted soil, while the least powerful ones are best suited to preparing previously cultivated soils. Depending on the source and size, these tractors cost between US $200 and $1500 [39].

6.2 Mechanized weeder

Two basic types of power weeders are available; ones that are worn on the operators back where weeds are cut and buried to shallow depths through arm movement, and others that resemble small walking cultivators that pass between crop rows (i.e., mini-cultivators). These machines require skillful use and maintenance so that crops are not injured and equipment remains in good working order. Using mechanized weeders, it is possible for a single operator to greatly reduce weeds from 0.5 to 1.0 ha per day. These units are available at a cost of US $200 to $380 each, depending on the size of the engine and the number of different attachments included with the purchase [39].

6.3 Power sprayers

This equipment is used to apply herbicides, pesticides, and fertilizers to crops. Sprayers range in size from portable units, to towed sprayers that are connected to a tractor, to self-propelled units with boom mounts many meters in length. Backpack sprayers are extremely useful in agricultural applications. These machines are easy to carry and use, but require that they be used with skill and caution. Power backpack sprayers cover about 0.5 ha per hour and cost about US $250, with larger and more expensive trolley options also available [39].

6.4 Earth augers

Also called post-hole diggers, this machine can be used to make planting pits and basins that collect rainwater. It consists of an engine powering a vertical shaft that rotates screw blades to displace soil, resulting in a cylindrical hole. A skilled operator can prepare a 40 cm deep hole of 25 to 30 cm wide in less than a minute, much faster and less strenuous than excavation by hand. Commercially available earth augers sell for about US $230, including an assortment of screw blades [39].

6.5 Drip irrigation

This system slowly delivers water onto the roots of plants in a way that strategically places moisture and minimizes evaporation. Drip irrigation distributes water through a network of valves, pipes, tubing, and emitters that operate at a relatively low water pressure. While complicated in design and expensive to install, the savings in water and yield improvement are substantial. One hectare of land can be placed under drip irrigation for about US $2400 [39]. The technology is typically associated with horticulture production but is also applicable to staple crops.

6.6 Rain guns

Imitating natural precipitation, this technology involves extended "water throw" through a sprinkler. Rain guns can cover large areas of field from relatively few stations and are portable which allow them to be moved between field locations according to needs and schedules. These guns require high water pressure and flow and can project water for up to 60 m in distance, covering a circular area of 1.1 ha. Its coverage is adjustable in terms of distance, droplet size and completeness of circular angle, with 360° projection most common. A complete rain gun irrigation system, including a high pressure pump, may be purchased for about US $670 [39].

6.7 Mechanized threshers

Unlike traditional hand separation of seed and grain from crops, this power equipment is more time-efficient and less laborious. Thresher machines are powered by small engines and consist of a feed chute that results to a spinning drum where crop residues are separated from seeds, and then a blower that removes lighter chaff. Operators put in dried harvest materials and can process between 150 and 500 kg of product per hour. Multi-crop threshers that handle numerous crops or specific ones for maize are sold by suppliers for about US $780 [39].

6.8 Residue cutters and choppers

Motorized processing of stover is ideal for mixed crop-livestock farming, particularly where wastes are plentiful and underutilized. Small, motorized cutters are easy to move, allowing residue recovery from several hectares in a day. Use of motorized choppers and crushers make it possible to provide suitable feed and mulch for soil cover while saving time and effort. Depending on the model, throughput capacities range from 1 to 1.5 ton of stover per hour and cost between US $1000 and $1500 [39].

7. Value chains in African agricultural transformation

Major efforts are underway to transform African agriculture [7]. This goal is extremely complex [9] as it involves different value chain segments and the technologies they depend on [8]. This chapter describes these value chains in terms of key modernizing technologies and the manner they benefit approaches to efficiency, substitution, and redesign over time [4], focusing upon four distinct models of value chains supporting cassava, wheat, small ruminants, and small-scale farm machinery. Cassava illustrates a value chain that strengthens primarily through the process of substitution [4] in that it progresses through the iterative introduction of a stream of different technologies along an entire value chain over time. Earlier in this Chapter, we describe 10 key technologies critical to this progress that are positioned along the entire cassava value chain. A net effect of modernizing the value chain is a transition of cassava from a subsistence crop to an agro-industrial export commodity [11]. The substitution of wheat in the bakery industry by high quality cassava flour is critical to reduce wheat import across the continent. At the same time, the technologies are suited to a wide variety of farmers, allowing them to adopt what is suited to their household and socio-economic needs. Many other agricultural commodities follow this pattern such as rice, maize, and sweet potato, where new technologies are streaming toward agricultural input supply markets in response to improved agricultural production strategies.

Wheat is a different case in that its introduction is largely the result of the intended redesign of agricultural systems. Admittedly, wheat is grown in the African highlands, and it represents an important commodity there, but the demand for wheat products in Africa greatly exceeds what can be produced. This led to massive imports that burden many economies, and the best solution is to greatly expand wheat production to new, warmer areas. The recent availability of heat tolerant varieties opens this possibility but only in conjunction with other accompanying production inputs and management strategies. Many African countries are now looking at ways to grow these new varieties of wheat in places traditionally reserved for other cereals [20]. The possibility of producing irrigated wheat in the Sahel during the relatively brief cool and dry season will have profound effects upon entire economies, but to achieve this vision requires that entire technology toolkits be adopted [21]. Growing wheat in rotation with maize and sorghum in sub-humid areas raises similar opportunities. The importance of mechanized bed preparation, planting, and harvesting suggests that larger, commercial operations are better positioned to enter new wheat production. Conservation Agriculture practices and their focus upon residue management and herbicides reinforces this likelihood. The challenge is to design wheat production systems that do not exclude smaller-scale producers, and this is achieved in large part by the availability and packaging of key production inputs and the ability of households to work together through collective operation of equipment. In much the same way, soybean is a commodity requiring strong elements of production system redesign [4] and offering strong agro-industrial economic advantages.

Another critical agricultural value chain is that of goats and sheep, and the technologies that support its transition from pastoral subsistence to commercial enterprise. Great effort is devoted to planned improvement of these systems as they are often situated within fragile drylands. Most of this improvement is related to assuring reliable supply of feed and water and building the quality of animal herds with time. To a large extent, this improvement relies upon collective management, but many key products related to health and feeding of animals become increasingly important. An important gain becomes disease prevention through veterinary products and services. With time, animals progressively move toward greater care by their managers in terms of protective confinement, feeding and health protection. Intensified management results in faster weight gains and improved meat quality. Processing crop residues into feeds, and supplementing them with blended grains, vitamins and minerals becomes a separate economic activity, as does short term fattening, resulting in greater differentiation along the value chain. This in turn results to de-risking of these value chains, leading to more sedentary lifestyles and higher standards of living.

Small-scale farm machinery is a distinctive value chain. Power equipment are not a commodity in themselves, but rather provide labor saving services along the production and processing value chain that are critical to the modernization of small-scale farming operations across Africa. Virtually every task currently performed by hand labor may be accomplished with greater ease by machine; be it tillage, weeding, planting, pest control, threshing and chopping. Note that these tasks are spread along the value chain, suggesting that these services are performed by different machines throughout the season. The ease of machine operations is extremely important to re-attracting youth to career paths in agriculture, a crucial component to the larger agricultural transformation equation. The challenge before the agricultural development community is how to make these machines more available and affordable to less affluent farmers, suggesting that collective ownership or service contracting may be in order. Many of these less affluent farmers serve as laborers on other farms, delaying the preparation of their own lands. Access to machinery helps to resolve this inequity.

The proportions of different modernizing technologies across key commodities reveal the importance of balance between new and better varieties and breeds, production inputs, improved management practice and agro-processing innovations (**Table 1**). In large part, crop improvement is a lead innovation, but without good use of accompanying production inputs and management practices, the returns to investment from improved germplasm is reduced. At the same time, agricultural intensification requires that value be added to crop surpluses, and these require processing innovations. A weakness is found in the inconsistency of digital tools, with some commodities supported by relevant tools and others not. Across all six commodities under discussion, the distribution of modernizing technologies in support of value chains in sub-Sahara Africa is Production Input Product > Genetic Innovation Product > Agricultural Management Practice > Agro-processing Innovation > Digital Support Tool.

A unique framework to assess yield improvement for different commodities undergoing agricultural transformation in Africa is presented in **Table 2**. In this case, the nature and cost of improvements are compared to the resultant yield increase to

Commodity	Genetic innovation product	Production input product	Agricultural management practice	Digital tool	Agro-processing Innovation
			f		
Cassava	0.25	0.25	0.08	0.17	0.25
Rice	0.30	0.30	0.10	0.10	0.20
Maize	0.30	0.50	0.10	0.10	0.00
Wheat	0.30	0.20	0.40	0.00	0.10
Soybean	0.20	0.30	0.20	0.00	0.30
Goats & sheep	0.10	0.30	0.30	0.00	0.30
All commodities	0.24	0.31	0.20	0.06	0.19
Farm machinery	0.00	1.00	0.00	0.00	0.00

Table 1.
The proportion of priority modernizing agricultural technologies based on commodity and type.

Commodity	Improvement	Increased cost (US$ per ha)	Yield increase (ton per ha)	Cost per ton increase (US$ per ton)
Cassava	improved cuttings, increased fertilizer, mechanized harvest	$614	23.4	$26
Maize	additional N topdressing, additional grain bags	$76	0.70	$109
Rice	irrigation and additional fertilizer, additional grain bags	$159	1.20	$133
Wheat	adapted seed, additional fertilizer, mechanization, irrigation, additional grain bags	$440	2.47	$178
Soybean	inoculant, blended fertilizer, herbicide, additional grain bags	$91	0.62	$147

Table 2.
Approaches and costs of yield improvement, resultant increase and the cost per ton of yield increase for five key commodities in Sub-Saharan Africa.

calculate the cost per ton of yield increase. In these scenarios, cassava requires considerable investment but results in considerable yield improvement, in part because the harvest of tubers is relatively high in moisture. This investment requires introduction of 10,000 cuttings of improved cuttings per ha, application of an additional 200 kg per ha fertilizer and conversion to mechanized lifting of tubers but is more than offset by massive increase in tuber yield. Cereals benefit from nitrogen topdressing (maize), irrigation and fertilizer (rice) and system redesign (wheat), resulting in yield increases ranging from US $109 to $178 per ton. Considering the world price of these commodities and the need for their massive importation to Africa, intensification of cereal production through strengthening value chains remains an important developmental objective. Substantial increases of soybean yields result from combining legume inoculant, blended fertilizers and pre-emergence herbicides, all production inputs that require the establishment of input manufacturing capacity and expanded delivery capacity. Note that both wheat and soybean require considerable redesign of current cropping systems as these are for the most part introduced crops in areas of expanded production, and not well known to producers.

8. Positioning technologies within development projects

As small-scale farmers continue to move toward commercial farming practices, their level of engagement with value chains continues to grow, and this is a unifying principle underlying agricultural transformation. This reliance also requires accompanying support to and from governments and successful integration of key technologies through value chains via development projects becomes a critical piece of the transformation equation. Projects awarded by multilateral donors such as the World Bank, African Development Bank and International Fund for Agricultural Development among others, provide loans and grants to countries to modernize agriculture as highlighted in **Figure 4**. Positioning the right mix of commodities and technologies within those projects becomes critical to their success, as is the meaningful inclusion of the various value chain actors depicted in **Figure 3**.

The Technologies for African Agricultural Transformation (TAAT) Project [8] led by IITA in collaboration with several other research and development organizations [9] serves as a mechanism to identify and promote proven modernizing technologies

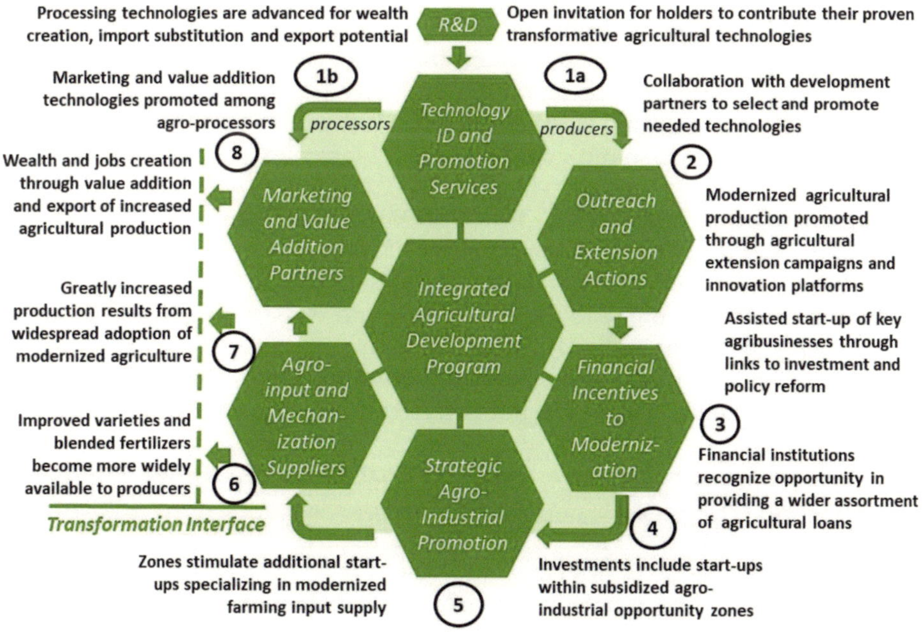

Figure 4.
A stepwise model of coordinated roll-out of modernizing agricultural technologies.

and to systematically match them to the projects of development organizations. This is a relatively new function of the CGIAR (formerly the Consultative Group for International Agricultural Research) as it seeks to work more effectively with national counterparts and Sub-Regional Organizations [41]. These development projects must engage and benefit three parties: rural communities, the private sector and government agencies [42]. Their design and implementation requires problem-solving and strong alliances with clear agreement on which difficulties exist and how best to resolve them, and TAAT offers growing expertise in this area. Examples of how TAAT has contributed to the design of several value chain interventions within recent projects follows (**Table 3**).

8.1 Multi-sectoral Approach for Stunting Reduction Project (MASReP)

This Ethiopian project advances agricultural technologies related to value addition to sweet potato with particular focus upon pregnant and lactating women and children under 5 years old.

8.2 Projet d'Appui au Programme Graine Phase 1 (PAPG1)

This project in Gabon focuses upon multiplication systems for cassava and plantain, including the capacity development of national partners. This effort includes the rehabilitation of two laboratories for SAH production (see Section 3.5) and building capacities in developing and multiplying new cassava varieties [12].

8.3 Agricultural Markets, Value Addition and Trade Development Project (AMVAT)

AMVAT addresses fundamental constraints to agricultural growth in South Sudan by taking an agricultural value chain approach in support of maize, sorghum, groundnuts, and sesame. The project works through 20 aggregation centers, 100 producer associations and 10 seed enterprise groups. It also provides food safety training. TAAT assists with the maize and sorghum value chains.

8.4 Program to Build Resilience for Food and Nutrition Security in the Horn of Africa (BREFONS)

This regional project operates across several countries (Djibouti, Kenya, Somalia, South Sudan, and Sudan) and select value chains (maize, millet and sorghum, livestock). Technologies holding commercial potential include climate-smart maize and bio-fortified sorghum and pearl millet varieties.

8.5 Le Projet d'Appui au Développement des Chaines de valeurs Agricoles dans les Savanes (PADECAS)

This project in Central African Republic focuses upon bean and cassava commodity value chains. Commercially viable technologies for beans include production of bio-fortified varieties, seed coating with agro-chemicals, wider distribution of herbicides, distribution of hermetic bags for safer grain storage, and milling of beans for use in blended flours. The cassava value chain multiplies plantlets and cuttings; relying upon imported agrochemicals for pest and weed management [12].

Development project	Investment	Commodities	Technologies	Comment
Multi-sectoral Approach for Stunting Reduction Project (MASReP)	$31 million over 4 years (Ethiopia)	1	3	Sweet potato value addition
Projet d'Appui au Programme Graine Phase 1 (PAPG1)	$106 million over 6 years (Gabon)	2	4	Cassava and plantain multiplication
Agricultural Markets, Value Addition and Trade Development Project (AMVAT)	$10 million over 4 years (South Sudan)	2	5	Focus upon value addition
Program to Build Resilience For Food And Nutrition Security In The Horn Of Africa (BREFONS)	$138 million over 5 years (Djibouti, Kenya, Somali, South Sudan)	4	7	Climate-smart and pest management of dryland commodities
Le Projet d'Appui au Développement des Chaines de valeurs Agricoles dans les Savanes en RCA (PADECAS)	$11 million over 7 years (Central African Republic)	4	7	Cassava, rice, bean, livestock
Projet d'Appui au Developpement Integre de l'Economie Rurale (PROADER)	$28 million over 6 years (DRC)	3	9	Not including five target perennial cash crops
Programme De Développement De La Zone Spéciale De Transformation Agro-Industrielle De Ngandajika (PRODAN)	$70 million over 5.7 years (DRC)	4	19	Agro-industrial park pilot operation in DRC
Cabinda Province Agriculture Value Chains Development Project (CPAVCDP) in Cabinda - Angola	$101 million over 6 years (Angola)	5	29	Cassava, maize, aquaculture, livestock, cowpea, soybean
Programme intégré de développement et d'adaptation au changement climatique dans le bassin du Niger (PIDACC)	$283 million over 6.5 years (9 countries)	3	25	Rice, maize, wheat
DRC Agricultural Transformation Agenda (DRC-ATA)	$24 million over 2 years (DRC)	6	40	Nationwide agricultural transformation program
Total (Mean ± SEM)	$803 million	(3.2 ± 0.5)	(14.8 ± 4.0)	Average 4.6 technologies per commodity

Table 3.
Selected agricultural development projects in Africa and their number of component commodities and technologies.

8.6 Projet d'Appui au Developpement Integre de l'Economie Rurale (PROADER)

PROADER promotes a dynamic and prosperous rural economy by improving agricultural and rural socioeconomic services, and by diversifying and enhancing agricultural production. It operates in seven Provinces of DRC by providing agricultural extension services and inputs, as well as adding value through agro-processing.

8.7 Programme De Développement De La Zone Spéciale De Transformation Agro-Industrielle De Ngandajika (PRODAN)

PRODAN is a pilot operation aimed at implementing national policies and strategies for the development of agro-industrial parks in 22 provinces. It is intended to stimulate growth in the agricultural sector, ensure food and nutritional security for the surrounding population, and generating sustainable jobs and income.

8.8 Cabinda Province Agriculture Value Chains Development Project (CPAVCDP)

This Angolan project leverages upon 29 best bet technologies across six key commodities (cassava, maize, cowpea, soybean, aquaculture, and livestock) by promoting the introduction of new crop varieties, seed certification, and improved farm management practices.

8.9 Programme intégré de développement et d'adaptation au changement climatique dans le bassin du Niger (PIDACC)

This project covers nine countries in the Niger River Basin (Benin, Burkina Faso, Cameroon, Cote D'Ivoire, Guinea, Mali, Niger, Nigeria, and Chad) and promotes 25 technologies related to rice, maize, and wheat focused upon training of extension workers and farming communities.

8.10 Agricultural Transformation Agenda in DR Congo (ATA-DRC)

This project includes cassava, rice, maize, bean, wheat and aquaculture value chains. Through the project, improved crop varieties were introduced through a combined approach to seed systems improvement that includes large state-run farms, local entrepreneurs, and international seed companies. The project also includes mechanization, agro-processing and youth empowerment activities at multiple locations. Its immediate goal is to reduce dependence upon wheat imports (see Sections 3 and 4).

Ten development projects appear in **Table 3** to describe their number of commodities and technologies they support. Overall, these projects represent an investment of US \$803 million (mean \$80.3 ± SEM 26.6 million) resulting from 21 country agreements supporting 32 commodity value chains and 148 technologies. Project participation ranged between one and nine countries (mean 2.1 ± 0.8) and project durations varied between two and seven years (mean 5.2 ± 0.5 years). Each project supported an average 3.3 commodities (SEM ± 0.5) and 14.8 technologies (± 4.0). When the number of projects, countries and duration are considered, each country receives \$8.7 ± 1.8 million per year (calculated from **Table 3**). Under these same conditions, each commodity receives \$5.0 ± 1.4 million per country/year and each technology receives \$1.5 ± 0.5 million per country/year to fund its value chain support. Together these projects represent a massive developmental investment that is ultimately intended to modernize the value chains of key food commodities.

These agricultural development projects follow a variety of approaches in terms of the number of commodities and technologies they advance, and the relative importance of capacity development between grassroots and private sector efforts that support them. The design of these projects and their level of funding depend upon their objectives, but there appears to be a limit to the number of technologies per commodity value chain (=4.6 calculated from **Table 3**) any given project can advance within a developmental context. This number of accompanying technologies is considerably less than the number of "toolkit technologies" considered essential to agricultural modernization (see Sections 3 to 5). In that case, careful consideration and balance must be applied to the selected technologies within any given project, too often limited to a few production inputs, or better yet the projects themselves could be designed in a more comprehensive manner by supporting modernizing technologies along the entire commodity value chains. In fairness, many of these projects are designed from a humanitarian rather than a value chain perspective and their component technologies fully emerge only as they are implemented, suggesting an important backstopping role for projects such as TAAT [8, 9] and organizations such as the CGIAR [41].

9. A coordinated approach for rollout of modernized agriculture technologies

Our promotion and scaling model operates upon the assumption that African small-scale farmers seek to acquire the best crop varieties and animal breeds for their production and market conditions, raise them with recommended accompanying inputs, conduct their production through sound management advice, and do so in a labor- and cost-effective manner [8]. In this way, agricultural transformation is the process that leads to increased farm productivity, making farming commercially viable and strengthening linkages with other sectors of the economy, particularly agro-processors. This responsibility requires more than simply promoting individual agricultural technologies within value chains; rather it advances suites of proven input products, production practices and conducive policies in a coordinated manner as transferable developmental assets. These Transferable Assets are proven innovations that address known constraints and have high potential for scaling across a wide range of settings and form the basis for an Integrated Agricultural Development Program (**Figure 4**).

Agricultural transformation is carefully designed and sequential. Many needed technologies are identified through past agricultural research and development efforts, and more are in the process of refinement (**Figure 4**). Within the context of an Integrated Agricultural Development Program, a call is issued to technology holders across the research and development community, inviting their participation and technical contributions. These include both production (**Figure 4(1a)**) and value-added processing technologies (**Figure 4(1b)**) positioned along the entire value chain (**Figures 1** and 2). Those production technologies deemed proven and ready for scaling [43] are promoted through agricultural extension campaigns and stakeholder innovation platforms (**Figure 4(2)**). Those technologies deemed most worthy of investment are offered incentives through assisted agribusiness expansion or startup (**Figure 4(3)**). In many cases, this involves strengthened placement within value chains through strategic investments within development project activities. As these businesses gain recognition and clients, banking institutions are expected to offer further opportunity for investment through an increasing number and volume of financial instruments and ventures [44].

Providing incentives for and directing the course of that investment is a major responsibility of the Integrated Agricultural Development Program as a means of agricultural transformation (**Figure 4(4)**). A critical next step is the strategic establishment of agro-industrial centers, particularly those that manufacture and distribute production inputs and equipment necessary to modernized agriculture. These Centers must be flexible and scalable, but in general they include seed multiplication and processing, fertilizer blending and packaging, feed blending for different animal enterprises, and the fabrication and distribution of machinery needed to eliminate the drudgery associated with small-scale farming (**Figure 4(5)**). These Centers may range in scope and size from modest agribusiness incubations promoting relatively few enterprises and technologies [45] to colossal Special Agro-Industrial Processing Zones supported by a consortium of development institutions [46] as either serves the same modernizing purpose. They not only function to increase the availability of proven production inputs and equipment, but also to train others in required skills so that parts of the Center may be propelled by the private sector, leading to a proliferation of Agro-Input and Mechanization Suppliers. In this way, the Centers operate as agribusiness incubators and are particularly attractive to youth empowerment efforts.

At this point, an agricultural transformation interface emerges. Improved varieties and blended fertilizers become more widely available to producers (**Figure 4(6)**). Greatly increased production results from widespread adoption of modernized agriculture (**Figure 4(7)**). Wealth and jobs are created through value addition and export of increased agricultural production (**Figure 4(8)**). Value chains are positioned to more positively interact with the larger agricultural and financial communities to transform agricultural production in alignment with large national programs. The immediate effect is to greatly increase production, so it is important to anticipate marketing requirements [47]. This anticipation involves the establishment of commodity collection points, the distribution of packaging materials, systems of quality assurance and the scaling of agro-industrial capacities. The impacts of successful application are measurable over the near- and mid-term. An immediate effect among smallholder communities is increased food supply and diversity. A mid-term goal is the reduction of food imports and an increase in agricultural exports, both leading to an increase in foreign reserves and the creation of decent jobs within the agricultural sector. All these achievements are reflected in stronger and more diverse commodity value chains.

Finally, a critical element within an Integrated Agricultural Development Program is the ability and need to undertake agricultural policy reform, particularly as it relates to investment in modernizing agricultural technologies by the private sector and client farmers. Opportunities for policy interventions stimulating national investment appear throughout this model (**Figure 4**) without being specifically recognized within this narrative. Strong political will for reform should not be equated with strong government intervention, as the most effective course of action is often to allow for greater commercialization through reduced government intervention; albeit within an enabling policy environment [7]. It is particularly important that governments do not excessively tax their early transformational successes [48]. This does not suggest that there are not many needed reforms; particularly those that open the seed and fertilizer sectors to investment, increase the movement of agricultural inputs and equipment into and between countries, and in many cases provide the needed financing to strengthen rural infrastructure. Instead, it implies the need for more strategic thought by governments to implement the right policies at the national and regional levels to support agriculture transformation.

10. Conclusions

This Chapter describes value chains from a technology perspective. This approach is particularly valuable within the context of development where agricultural systems are expected to modernize and transform and be able to respond to the economic and climate shocks we are seeing today. This is the case across all three transitional stages leading to sustainable management of agricultural systems involving: (1) greater production efficiency with current resources, (2) strategic substitution with more potent production inputs and managements, and (3) the redesign of systems that allows previously unobtainable types and levels of production.

Distinctions were drawn between the cassava commodity value chain, where current systems are steadily improved and commercialized, and wheat production in new production areas across Africa. In the latter case, excessive economic reliance upon imported wheat creates an economic burden that must be relieved. Similar types of technologies are required across both of these commodity value chains, but the manner of their integration varies within the scope of systems redesign; and in this case is much greater for the introduction of wheat as a product of a new and expanding cropping system. Interestingly, cassava flour resulting from agro-industrial processing of raw tubers can partially substitute for wheat flour; whether that wheat is domestically produced or imported.

Distinctions were also drawn between crop and animal enterprise value chains, in this case the intensification of goat and sheep production. The proportions of technologies as they are divided among genetic innovation products, production input products, agricultural management practice, digital agricultural tool and agro-processing innovation provides a technological signature useful to agricultural development strategies. The need for greater reliance upon mechanization, particularly the use of small-scale farming equipment as a means to reduce unacceptable levels of drudgery, was raised and a wide selection of these machines were identified for use across entire agricultural value chains.

Agricultural development projects and the interests that foster them seek to achieve agricultural transformation across Africa, and this involves the selection, promotion, and scaling of modernizing technologies. These intentions result from the resolve to both secure the continent's food self-sufficiency and to better propel its agro-industries. These projects are often based upon projected increases in production founded upon the potency and adoption of incrementally improved management technologies and products. This paper presents such modernizing technologies for cassava, wheat, and small ruminants as examples, but many other interventions are known [9, 42]. The African Development Bank recently identified the need for an additional 70 million tons of cereal grain (wheat, maize, rice and sorghum), 30 million tons of fresh cassava (= 10 million tons of cassava flour), 10 million tons of soybean, and 5 million tons each of domestic animals (poultry, beef, sheep, goats and swine) and fish (in aquacultural systems) to secure continent-wide food and nutritional security and to fully launch its agro-industries upon the world stage [49]. Agricultural value chains and the technologies they rely upon and distribute occupy a critical role in securing this target.

Acknowledgements

Information on the technologies described in this paper was provided by current or past TAAT Compact Leaders Adebayo Abass (TAAT Cassava Compact), Zewdie Bishaw (TAAT Wheat Compact) and Adeniyi Adediran (TAAT Livestock Compact). Charles Kibiru of Ikonic Farm Machinery (Kenya) provided information on small-scale farming equipment. Kennedy M. Kago and Dêêdi Sogbohossou of the TAAT Clearinghouse assisted in the compilation of African Development Bank project information. Nteranya Sanginga of the African Agricultural Leadership Institute provided insights into technology rollout and delivery. The IITA TAAT Clearinghouse is supported through a project of the Bill and Melinda Gates Foundation and the accompanying TAAT Program is funded by the African Development Fund of the African Development Bank. The contributions of all these individuals and organizations are gratefully acknowledged.

Author details

Paul L. Woomer[1,2], Rachel M. Zozo[1], Sabra Lewis[1*] and Dries Roobroeck[1]

1 International Institute of Tropical Agriculture (IITA), Nairobi, Kenya

2 African Agricultural Leadership Institute, Nairobi, Kenya

*Address all correspondence to: s.lewis@cgiar.org

References

[1] Porter ME. Competitive Advantage: Creating and Sustaining Superior Performance. New York, USA: Free Press; 1998. ISBN 978-0684841465

[2] Kaplinsky R, Morris M. A Handbook for Value Chain Research. Canada: International Development Research Center; 2012. Available from: https://hdl.handle.net/10568/24923

[3] Webber CM, Labaste P. Building competitiveness in Africa's agriculture: A guide to value chain concepts and applications. Agricultural and Rural Development. Washington D.C.: World Bank; 2009. 187 p. DOI: 10.1596/978-0-8213-7952-3

[4] Hill S. Redesigning the food system for sustainability. Alternatives: Global, Local, Political. 1985;**12**:32-36

[5] Hill S. Considerations for enabling the ecological redesign of organic and conventional agriculture: A social ecology and psychosocial perspective. In: Bellon S, Penvern S, editors. Organic Farming, Prototype for Sustainable Agricultures. Dordrecht, Netherlands: Springer; 2014. pp. 401-422. DOI: 10.1007/978-94-007-7927-3_22

[6] Pretty J, Benton TG, Bharucha ZP, Dicks LV, Butler Flora C, Godfray HCJ, et al. Global assessment of agricultural system redesign for sustainable intensification. Nature Sustainability. 2018;**1**:441-446. DOI: 10.1038/s41893-018-0114-0

[7] African Development Bank. Feed Africa: Strategy for Agricultural Transformation in Africa 2016-2025. Abidjan, Cote d' Ivoire: African Development Bank. 2016. 79 p

[8] Woomer PL, Mulei WM, Zozo R. A new paradigm in the delivery of modernizing agricultural technologies across Africa. In: Technology in Agriculture. London: IntechOpen; 2021. 23 p. DOI: 10.5772/intechopen.98940

[9] TAAT. TAAT in 2018: Preparing for African Agricultural Transformation. Ibadan, Nigeria: International Institute of Tropical Agriculture; 2019. 65 p

[10] Pretty J, Toulmin C, Williams S. Sustainable intensification in African agriculture. International Journal Agricultural Sustainability. 2011;**9**:5-24. DOI: 10.3763/ijas.2010.0583

[11] Lobenstein G. Plant virus diseases: Economic aspects. In: Mahy B, Van Regenmortel M, eds. Encyclopedia of Virology, 3rd ed. Cambridge, MA, USA: Academic Press; 2008. p. 197-201. DOI: 10.1016/B978-012374410-4.00693-2

[12] TAAT Clearinghouse. Cassava Technology Toolkit Catalogue. Clearinghouse Technical Report Series 006. Cotonou, Benin: Technologies for African Agricultural Transformation Clearinghouse Office, IITA; 2021. 28 p. DOI: 10.21955/gatesopenres.1116798.1

[13] Legg JP. African cassava mosaic disease. In: Mahy B, Van Regenmortel M, eds. Encyclopedia of Virology. 3rd ed. Cambridge, MA, USA: Academic Press; 2008. p. 30-36. DOI: 10.1016/B978-012374410-4.00693-2

[14] Nweke FI, Spencer DSC, Lynam JK. The Cassava Transformation: Africa's Best Kept Secret. East Lansing, MI, USA: Michigan State University Press; 2002. 273 p. ISBN: 9780870136023

[15] Sanoussi AF, Loko LY, Ahissou H, Adjahi AK, Orobiyi A, Agré AP, et al. Diversity, physicochemical and

technological characterization of elite cassava (*Manihot esculenta* Crantz) cultivars of Bantè, a District of Central Benin. The Scientific World Journal. 2015:674201. DOI: 10.1155/2015/674201

[16] Ayetigbo O, Latif S, Abass A, Müller J. Comparing characteristics of root, flour and starch of biofortified yellow-flesh and white-flesh cassava variants, and sustainability considerations: A review. Sustainability. 2018;**10**:3089. DOI: 10.3390/su10093089

[17] Kintché K, Hauser S, Mahungu NM, Ndonda A, Lukombo S, Nhamo N, et al. Cassava yield loss in farmer fields was mainly caused by low soil fertility and suboptimal management practices in two provinces of the Democratic Republic of Congo. European Journal of Agronomy. 2017;**89**:107-123. DOI: 10.1016/j.eja.2017.06.011

[18] Morgan NK, Choct M. Cassava: Nutrient composition and nutritive value in poultry diets. Animal Nutrition. 2016;**2**:253-261. DOI: 10.1016/j.aninu.2016.08.010

[19] Iwe MO, Michael N, Madu NE, Obasi NE, Onwuka GI, Nwabueze TU, et al. Physicochemical and pasting properties High Quality Cassava Flour (HQCF) and wheat flour blends. Agrotechnology. 2017;**6**:100-167. DOI: 10.4172/2168-9881.1000167

[20] Negassa A, Shiferaw B, Koo J, Sonder K, Smale M, Braun HJ, et al. The Potential for Wheat Production in Africa: Analysis of Biophysical Suitability and Economic Profitability. D.F. Mexico, Mexico: CIMMYT; 2013. 64 p. ISBN: 978-607-8263-28-8

[21] TAAT Clearinghouse. Wheat Technology Toolkit Catalogue. Clearinghouse Technical Report Series 011. Nairobi, Kenya, IITA: Technologies for African Agricultural Transformation, Clearinghouse Office; 2021. 32 p

[22] Ni Z, Li H, Zhao Y, Peng H, Hu Z, Xin M, et al. Genetic improvement of heat tolerance in wheat: Recent progress in understanding the underlying molecular mechanisms. The Crop Journal. 2018;**6**: 32-41. DOI: 10.1016/j.cj.2017.09.005

[23] Mengesha GG. Management of yellow rust (*Puccinia striiformis* f.sp. *tritici*) and stem rust (*Puccinia graminis* f.sp *tritici*) of bread wheat through host resistance and fungicide application in Southern Ethiopia. Cogent Food & Agriculture. 2020;**6**:1739493. DOI: 10.1080/23311932.2020.1739493

[24] Abeyo B, Badebo A, Gebre D, Listman M, editors. Achievements in Fast-Track Variety Testing, Seed Multiplication and Scaling of Rust Resistant Varieties: Lessons from the Wheat Seed Scaling Project, Ethiopia. D.F. Mexico, Mexico: CIMMYT; 2020. 106 p. ISBN: 978-99944-78-80-4

[25] Schmid RB, Knutson A, Giles KL, McCornack BP. Hessian Fly (*Diptera: Cecidomyiidae*) biology and management in wheat. Journal of Integrated Pest Management. 2018;**9**:1-12. DOI: 10.1093/jipm/pmy008

[26] Nsarellah N, Amri A, Nachit MM, Bouhssini ME, Lhaloui S, Lorenzoni C. New durum wheat with Hessian fly resistance from *Triticum araraticum* and *T. carthlicum* in Morocco. Plant Breeding. 2003;**122**:435-437. DOI: 10.1046/j.1439-0523.2003.00871.x

[27] Tidiane Sall A, Chiari T, Legesse W, Seid-Ahmed K, Ortiz R, van Ginkel M, et al. Durum wheat (*Triticum durum* Desf.): Origin, cultivation and potential expansion in Sub-Saharan Africa. Agronomy. 2019;**9**:263. DOI: 10.3390/agronomy9050263

[28] Tewabe D, Abebe A, Enyew A, Tsige A. Determination of bed width on raised bed irrigation technique of wheat at Koga and Rib Irrigation Projects, Northwest Ethiopia. Cogen Food & Agriculture. 2020;**6**:1712767. DOI: 10.1080/23311932.2020.1712767

[29] Landis DA, Saidov N, Jaliov A, El Bouhssini M, Kennelly M, Bahlai C, et al. Demonstration of an integrated Pest management program for wheat in Tajikistan. Journal of Integrated Pest Management. 2016;7:1-9. DOI: 10.1093/jipm/pmw010

[30] Olt J, Küüt K, Ilves R, Küüt A. Assessment of the harvesting costs of different combine harvester fleets. Research in Agricultural Engineering. 2019;**65**:25-32. DOI: 10.17221/98/2017-RAE

[31] TAAT Clearinghouse. Goat and Sheep Technology Toolkit Catalogue. Clearinghouse Technical Report Series 017. Cotonou, Benin: Technologies for African Agricultural Transformation, Clearinghouse Office, IITA; 2022. 32 p. DOI: 10.21955/gatesopenres.1116823.1

[32] Gizaw S, Getachew T, Edea Z, Mirkena T, Duguma G, Tibbo M, et al. Characterization of Indigenous Breeding Strategies of the Sheep Farming Communities of Ethiopia: A Basis for Designing Community-based Breeding Programs. Aleppo, Syria: International Center for Agricultural Research in the Dry Areas (ICARDA); 2013. Available from: https://hdl.handle.net/20.500.11766/7796

[33] Liniger HP, Mekdaschi SR. Sustainable Rangeland Management in Sub-Saharan Africa–Guidelines to Good Practice. Washington D.C.: TerrAfrica; World Bank; 2019. 408 p. Available from: https://documents1.worldbank.org/curated/ru/237401561982571988/pdf/Sustainable-Rangeland-Management-in-Sub-Saharan-Africa-Guidelines-to-Good-Practice.pdf

[34] Amole TA, Ayantunde A, Dangoma A. Least-Cost Rations for Sheep Fattening: A Manual for Livestock Farmers and Extension Workers in the West African Sahel. Nairobi, Kenya: ILRI; 2016. Available from: https://hdl.handle.net/10568/72736

[35] Maas B, Chiuri W, Zozo R, Musale D, Metre T, Birachi E. Using the 'livestock ladder' as a means for poor crop-livestock farmers to exit poverty in Sud Kivu Province, Eastern DR Congo. Chapter 11, p. 145-155. In: Vanlauwe B, van Asten P, Blomme G. (eds.). Agro-Ecological Intensification of Agricultural Systems in the African Highlands. London, UK: Earthscan, Routledge; 2010. ISBN: 9780415532730

[36] Jensen HA, Dolberg F. A conceptual framework for using poultry as a tool in poverty alleviation. Livestock Research for Rural Development. 2003;**15**:39. Available from: http://www.lrrd.org/lrrd15/5/jens155.htm

[37] FAO. Recognizing Peste des Petits Ruminants: A Field Manual. Rome, Italy: Emergency Prevention System for Transboundary Animal and Plant Pests and Diseases Unit, Food and Agriculture Organization of the United Nations; 1999. 34 p. Available from: https://www.fao.org/3/x1703e/x1703e00.htm

[38] Clottey SJA. Manual for the Slaughter of Small Ruminants in Developing Countries. Animal Production and Health Paper 49. Rome, Italy: Food and Agriculture Organization of the United Nations; 1985. 123 p. Available from: https://www.fao.org/3/X6552E/X6552E00.htm#TOC

[39] TAAT Clearinghouse. Small-scale Farm Mechanization Catalogue.

Clearinghouse Technical Report Series 015. Cotonou, Benin: Technologies for African Agricultural Transformation, Clearinghouse Office, IITA; 2022. 24 p. DOI: 10.21955/gatesopenres.1116893.1

[40] Zozo R, Lewis S, Kago K, Opaluwah A, Adetunji O, Asiedu E, et al. Report of the Rice Outcome Case Study: Seed Production of Resilient Rice Varieties and the GEM Parboiling System in Cote d'Ivoire. Cotonou, Benin: TAAT MEL Working Document No. 004, IITA; 2020. 24 p

[41] CGIAR System Organization. CGIAR 2030 Research and Innovation Strategy. Montpellier, France: CGIAR System Organization. No year. 35 p. Available from: https://gspace.cgiar.org/bitstream/handle/10568/110918/OneCGIAR-Strategy.pdf

[42] Woomer PL, Roobroeck D, Alia DY. Blending climate action and rural development in Africa's Sahel. In: Ozcatalbas O, editor. Sustainable Rural Development. London, United Kingdom: IntechOpen; 2022. 25 p. Available from: https://www.intechopen.com/online-first/81496

[43] Schut M, Cadilhon J-J, Misiko M, Dror I. The state of innovation platforms in agricultural research for development. In: Dror I, Cadilhon J-J, Schut M, Misiko M, Maheshwari S, editors. Innovation Platforms for Agricultural Development: Evaluating the Mature Innovation Platforms Landscape. Earthscan-Routledge (New York) and International Livestock Research Institute (Nairobi); 2016. pp. 1-15

[44] FAO and AFRACA. Agricultural Value Chain Finance Innovations and Lessons: Case Studies in Africa. Rome, Italy: Food and Agriculture Organization of the United Nations (FAO) and African Rural and Agricultural Credit Association (AFRACA); 2020. DOI: 10.4060/ca6345en

[45] Von Kaufmann R and Ariho A. (eds.).Special Issue on Agribusiness Incubation in Africa. Agriculture for Development. 2019;**38**:1-65

[46] Adefeko A. Special Agro-industrial Processing Zones (SAPZ): A catalyst for Africa's industrial revolution. The Cable (Nigeria). 2021. Available from: https://www.thecable.ng/special-agro-industrial-processing-zones-sapz-a-catalyst-for-africas-industrial-revolution

[47] FAO. The Future of Food and Agriculture–Trends and Challenges. In: Food and Agriculture Organization of the United Nations (FAO). Rome, Italy; 2017. 163 p. Available from: https://www.fao.org/3/i6583e/i6583e.pdf

[48] Aksoy MA. African Agricultural Reforms: The Role of Consensus and Institutions. Directions in Development (Series). Washington, DC: The World Bank; 2012. 313 p. Available from: https://openknowledge.worldbank.org/handle/10986/9307

[49] AfDB. Dakar 2 Summit-Feed Africa: Food Sovereignty and Resilience. 25-27 January 2023, Dakar Senegal. African Development Bank; 2023. Available from: https://www.afdb.org/en/dakar-2-summit-feed-africa-food-sovereignty-and-resilience/about-dakar-2-summit

Chapter 5

Value Chain Analysis of the Street Food Enterprises in the Rural Towns of Vhembe District, Limpopo Province

Tjale Clopus Mahopo, Noxolo Cebisa Nesamvuni, Azwihangwisi E. Nesamvuni and Johan van Niekerk

Abstract

This study sought to analyse the value chain of foods sold by street food vendors in the Vhembe district town of Limpopo province. Data were collected using mixed methods, including structured questionnaires administered via face-to-face interviews. Analyses of the value chain activities and actors were conducted among the stages of street food production. Overall, the enterprise sold commonly consumed street foods such as porridge, chicken, beef, and local vegetables, targeting government employees, school children, and daily town visitors. The value chain analysis showed that street food activity in the Vhembe district included purchasing, storage, production, and consumption. The main actors involved in the value activities of the street food enterprise were inputs suppliers, formal and informal traders, transporters, local authorities, and customers. The profit margin calculation highlighted that street food vendors purchasing inputs from the informal vendors without any added expense could make more profit than using formal traders. Future policy interventions should promote value addition along the food sold in the street, providing cold storage facilities closer to street vendors' stalls in the towns to encourage continued and safe production. The most prioritised production constraints should be addressed, such as access to finance, government support, and managerial skills.

Keywords: demographics, operation characteristics, street food enterprises, street vendors, South Africa

1. Introduction

The street food trade is an ancient practice [1, 2] common in several countries as a source of income. It provides inexpensive meals accessible to the population and represents the culture of typical and local food [3]. The World Healfth Organization [4] defines street food as foods and beverages prepared and sold by vendors on streets and other public places for immediate consumption [4, 5]. This definition emphasizes the retail location on the street, with foods sold from pushcarts, bicycles, baskets or balance poles, or stalls that do not have permanent walls [6]. Millions of people depend on a wide variety of ready-to-eat foods and beverages sold and sometimes prepared in the street or public places. Charman et al. [7] conducted a microenterprise census of five working-class township settlements, demonstrating its solid economic basis in the trade of food, takeaways, and drink.

Moreover, informal micro-enterprises could address South Africa's unemployment challenges and stimulate economic growth in marginalized areas. Hence, it was noted that Maloney [8] could contribute through the business generation and transfer of skills and experience to informal workers. Despite some critical research on this topic, there remain significant knowledge gaps, especially on the value chain of street food that could create employment and transfer skills. Nevertheless, the increasingly recognised role of informal food services for food security, especially for poor urban dwellers [9], understanding activities, mechanisms, and the trade environment within urban townships remains limited [10].

Thus, street foods have important implications for consumers who enjoy these foods and street food handlers who handle and serve these foods [11]. However, increasing population, migration from rural to urban areas, an increase in the number of women engaging in gainful employment, inflation, and the breakdown of the joint family system have supported the growth of the street food sector.

Long distances to commute between workplace and home often compel individuals to eat food from commercial outlets. The significance of street foods in contributing to the food and nutrition security of workers residing away from home in small towns and cities is widely recognised.

Literature has noted a paucity of data on how the street food chain system is organized and operated for market nodes, governance, challenges, and food safety issues. Such information is crucial for understanding the sector, identifying growth opportunities, and supporting informal economy policies and programs. Along the street food value chain, problems such as poor infrastructure and lack of financial assistance hinder the possible benefits that the value chain actors should have attained. Therefore, there is a need to employ a value chain analysis (VAC) framework to understand the value chain of street foods in the Vhembe district [12, 13]. The value chain is a range of activities required to bring a product or service from conception through the different phases of production, transformation, and delivery to final consumers [14]. Then VCA seeks to understand how chain activities are organized, costs incurred, and benefits shared among chain participants. Its management is to add value and segment the market with differentiated products designed to increase profitability at all stages in the chain [14–17]. The value chain also deals with the institutional arrangement governing the activities, actors, their relationships, the linkages, and market prices in and out of each actor in the chain [18].

Porter [19] highlights that a firm's activities business unit is appropriate for constructing a value chain. Value chains are more complex in the real world and focus on systems and how inputs are changed into the outputs purchased by consumers [14]. In this study, we used the stages of street food used by Cortese et al. [16] (**Figure 1**), highlighting that a typical agricultural or food value chain consists of chain actors that transact a product as it moves through the value chain. These actors include input suppliers, farmers, traders, processors, transporters, wholesalers, retailers, and final consumers [16].

Limpopo province has the highest number of street food vendors, mainly concentrated in the Vhembe district ([20, 21]a). For this reason, Vhembe district was selected for this study to contribute to the body of knowledge on street food enterprise in the Vhembe district. Most street food vendors are found in public places such as taxi ranks

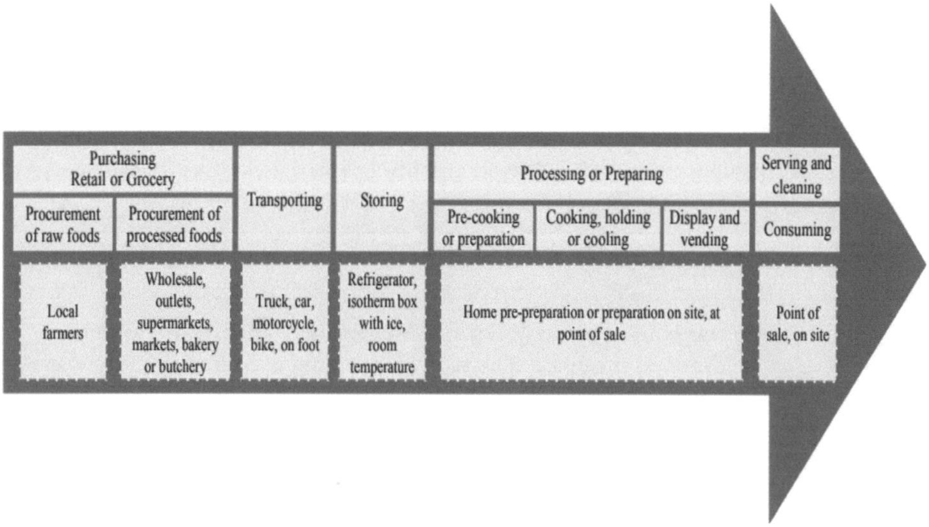

Figure 1.
Stages of the street food value chain. (Source: [16]).

and the road [22–24]. The Vhembe district draws special attention in the province because of a high increase in cooked street foods [20, 21, 25]. The main objective of this study was to map and analyse the value chain of street foods in the Vhembe district. The study seeks to identify the major players and the linkages in street food vendors' value chain. Moreover, we determined the chain's value-added and associated costs from purchasing to consumption of street food vendors.

To achieve this, we used the six-steps VCA process by Porter [19] to explain the value chain activities of street food enterprises. The following steps were used in this study to conduct a VAC of the street food vendors in the Vhembe district rural towns: firstly, in our previous work [26, 27], we described and defined the product sold by the street food vendors in the Vhembe district looking at the operational characteristics of the street food vendors (Step 1). This study sought to analyse the value chain of foods sold by the street food vendors in the Vhembe district town of the Limpopo province. This was done by identifying the prominent value chain actors, mapping significant steps (Channels), and analysing the profit margin share of the enterprise using four different types of plate sold.

2. Material and methods

2.1 Study area

The Vhembe District Municipality consists of four local municipalities: Makhado, Thulamela, Musina, and Collins Chabane. Botswana in the west, Zimbabwe in the north, and Mozambique share the borders. It covers an area of 21,402 km^2 of mostly rural land. About 32.65% of South Africans live in rural settlements [28]. Due to the high level of unemployment and poverty, inhabitants of the district have developed several survival strategies, including the street food enterprise of major trading points within the local municipalities identified in the towns of Makhado, Thohoyandou, and

Musina, respectively. The informal sector in South Africa contributes 8% of the country's GDP and supports 27% of all working people. The Vhembe District has a population of 14,02,779 people, increasing from 12,94,722 in 2011. According to Community Survey, in 2016, most of these people resided in Thulamela (4,97,237), followed by Makhado (4,16,728), Collins Chabane (3,47,974) and Musina (32,009). Thulamela LM is the most populated Municipality in the district, while Musina LM is the least populated Municipality (Limpopo Provincial department 2021).

2.2 Field survey: sampling

The field survey was conducted in Makhado, Musina, and Thohoyandou in the Vhembe District of Limpopo province. The type of sampling used in this study was non-probability sampling, as the research population is unknown [3, 29]. We used convenience sampling to select 511 street food vendors enterprises in different regions of the Vhembe district, Limpopo Province of South Africa. The researcher established and maintained a complete list of the primary unit component from the Vhembe district's municipalities. However, the municipalities had no records in place of street vendors. From the intended population of street vendors, the sample size was determined based on Yamane's formula of the year 1969 [30, 31]. The formula states that at the significance level of 90%, the minimum sample size for the unknown population is 100 [30, 31].

A total of 511 street vendors from Musina (168), Makhado (36), and Thohoyandou (306) towns participated in the study. Only street food vendors selling ready-to-eat or cooked food participated in the study. A convenient sampling technique was used considering street vendors' work. Street food vendors were obtained from taxi ranks, along the main roads, industrial areas, malls, schools, and lastly, between residences' streets, especially in Musina town.

2.3 Data collection

Data were collected by trained enumerators led by the research team using a structured questionnaire through face-to-face interviews. The questionnaire captured socio-demographic characteristics, including age, gender, and education. The quantitative data was collected from the street food vendors by asking them to estimate the proportion of value flow. Additional information was captured on production, input sources, storage, transportation mode, institutional support, and distribution of profit margin. Where possible, figures were obtained through consensus by achieving a majority. Qualitative data was collected by employing interview with 54 street food vendors who were available. Additional information generated was stakeholders and their roles, mapping the stages of street food production. During the sessions, we explored why street food vendors could not affiliate with an association.

3. Results

3.1 Introduction

The purpose of VCA seeks to understand how chain activities are organised, the cost incurred, and the benefits shared among chain participants. This section of the paper reports on the activities conducted to determine the value chain of the street

DOI: http://dx.doi.org/10.5772/intechopen.110398

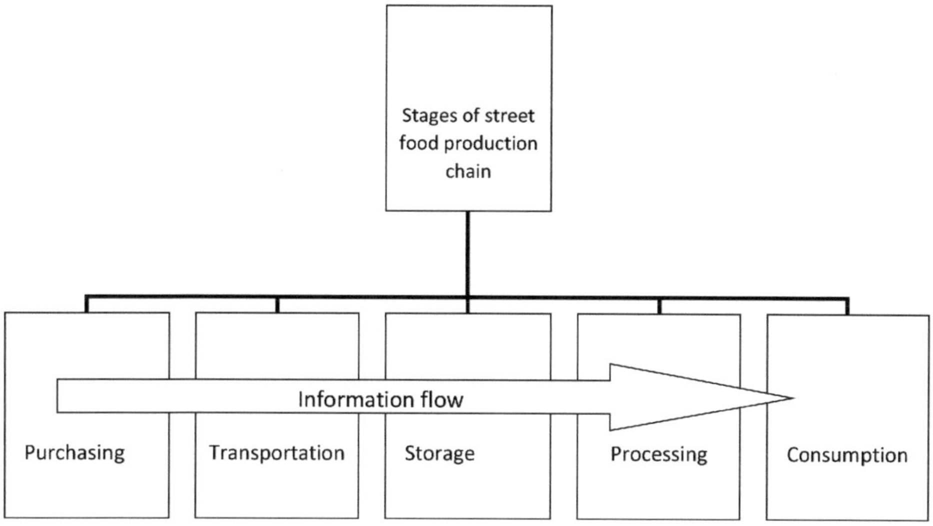

Figure 2.
Street food vendors food production chain activities.

food enterprises in the rural towns of the Vhembe district. A step-by-step method was used to analyse the street food enterprise. The following activities were conducted (1) identification of the marketing channels, (2) Mapping the value chain activities, (3) identification of the chain actors, (4) identifying value chain linkages, and lastly, (5) Determined the profit margin of street food vendors. The street food production chain activities in the Vhembe district are highlighted in **Figure 2**.

3.2 Description of the street food value chains

3.2.1 Marketing channels

Figure 2 shows the value chain channels identified amongst the street food vendors in the Vhembe district. The value chain activities of the street food vendors in the Vhembe District are undeveloped and straightforward, with no infrastructure. The main marketing channels identified are the input suppliers, transportation, storage, production, and consumption. The channels highlight that street food vendors have various food storage and processing places, from input suppliers to transportation to consumption. These highlights that the final product market of street food is consumers (government employees, school children).

3.3 Value chain analysis

3.3.1 Mapping the value chain activities

The value chain activities of the street food vendors start from purchasing, transportation, storage, processing, and lastly, consumption, depicting the flow value chain activities of street food vendors.

The street food vendors' value chain activities are explained in subsections 3.3.1.1–3.3.1.5.

3.3.1.1 Purchasing

Purchasing is the first stage of street food enterprises. During this stage, street food vendors purchase inputs from informal and formal stakeholders. The data showed that street food vendors purchase raw and processed products from informal traders such as bakkies/farmers/formal supermarkets. Raw products were reported to be mainly meat category were chicken—dressed cull layers constituted (88.8%) of the meat product purchased. Beef was reported at 68.3% with its associated Beef Offal at (14.3%).

From the vegetable category, the main raw product was Green Leafy Vegetables constituting (60.1%) purchased from hawkers using informal supply contracts. Maize meal by far was the most predominant raw product purchased (97.1%) from retailers because it happens to be a staple food in the towns, district, province, and the country at large.

Culled layers, and beef offal gives street vendors a competitive advantage since these products are a by-product being sold from major poultry and beef processing facilities [27]. The same applies to green leafy vegetables which are purchased form Bakkies or produced by street vendors in their back yards. Based on convenience the maize meal that is purchased gets augmented by by-products from maize processing facilities to create varieties, delicacies, and flavors (sour versus normal) of pap cooked the indigenous way. Also, these products mimic the indigenous and cultural meal that the customers mainly rural (82.2%), government employees (59.5%) and school children (41.5%) demand due to taste and value for money **Figure 3** [27].

3.3.1.2 Cost of inputs

The average monthly running cost of the inputs used to produce a plate of at least R45/30 of chicken or beef was estimated at R1800. On average, their daily cost of vegetables purchased from street food vendors or hawkers was R21.65, while vegetables from supermarkets cost R18.75–R51.00. Other food items purchased from the smallholder farmers included live chicken broilers with an estimated R180.83 daily. Street food vendors reported challenges such as a lack of continuous supply from street vendors/farmers and bakkie sellers; hence some of the inputs were purchased from retailers. On average, the cost of meat was between R188.00 and R279.00 from retail or supermarkets. [26] indicated the list of input suppliers that the street vendors in the study are used.

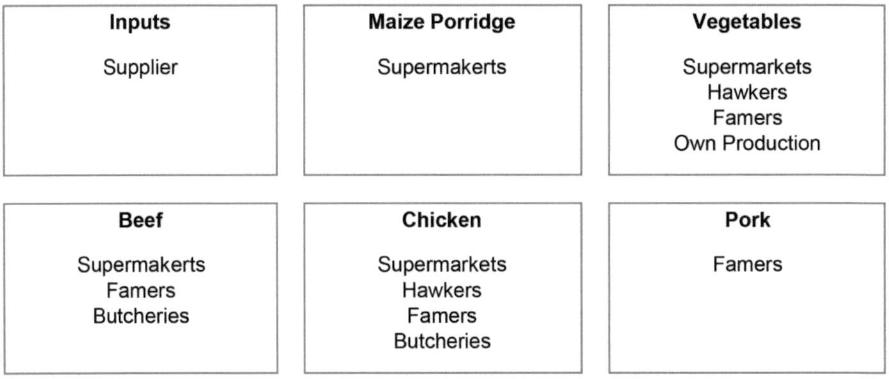

Figure 3.
Main food items purchased by street food vendors from chain stores and suppliers.

3.3.1.3 Transportation of ingredients and pre-prepared food

The second stage of street food value chain activities identified is transportation. The mode of transport, distance, and time spent acquiring inputs by the street vendors from suppliers are reported in **Table 1.** The average monthly transport cost for all street vendors who used any transport was estimated at R340.00. More than half of the street food vendors (54.6%) walk to get their input suppliers, while public transport, such as taxis and buses, was used by 33.3% and 8.4%, respectively. Only 1.6% of street food vendors reported that suppliers deliver inputs. Very few street food vendors used private cars (4.1%). The longest distance travelled was more than 10 km (4.5%). However, most reported walking less than one kilometre (6.5%) and taking close to 10 minutes (31.8%) to suppliers.

3.3.1.4 Storage

Table 2 presents the frequency and associated percentages of street food vendors' storage practices. The primary purpose of the storage is to extend the product availability over a more extended period than if it were sold immediately after harvest. In this study, most street food vendors reported having a place to store their business

	Frequency (n = 511)	% of participants	Median
Amount paid for transportation (R)			340.00
Mode of transportation of food			
Walk	279	54.6%	
Taxi	170	33.3%	
Bus	43	8.4%	
Car	21	4.1%	
Suppliers' delivery	8	1.6%	
Distance to inputs suppliers			
Less than 1 km	33	6.5%	
1–5 km	11	2.2%	
6–10 km	11	2.2%	
More than 10 km	23	4.5%	
Time to walk			
less than min walk	163	31.8%	
11–20 min walk	93	18.2%	
21 and more minute walk	25	4.9%	
Time to drive			
5–10 min drive	30	5.9%	
11–20 min drive	30	5.9%	
More than 20 min drive	75	14.7%	

Table 1.
Frequencies and associated percentages of mode of transportation and distance to inputs suppliers of ingredients and pre-prepared food for street food vendors.

Variables	Frequency	Percentage (%)
Availability of storage	*N* = 511	
Yes	357	70.0
No	154	30.0
Storage of stock	*N* = 357	
Home	97	27.4
Stall	75	21.2
Storeroom	108	30.5
Rented room	70	19.8
Other	4	1.1
Availability of access to on-site storage of processed food	(511)	
No	465	91.0
Refrigeration/freezer	46	9.0

Table 2.
Frequencies and associated percentages of street food vendor's storage practices.

materials including inputs (70%). Of those who reported having a place to store their enterprise material and inputs, 27.4% indicated that they store them at home, while 30.5% had rented rooms close to their stalls. Less than ten percent had access to storage facilities like freezers (3.5%) and refrigerators (5.3%) when selling at the stall. Less than 10% of the street food vendors had access to a freezer/refrigerator to store processed foods close to their enterprise, which risked food safety. An indictment to the local authority was the fact that 91% had no availability of access to on-site storage and no facility of safe refrigerators and freezers. This is mainly because the street vending business is not regulated nor managed within the developments plans of local authorities.

3.3.1.5 Production/processing

Food processing is a critical stage in the street food value chain. It entails pre-preparation and preparation of different food items, as shown in **Figure 2**. Pre-preparation is vital because different foods need to be prepared in different ways. In this stage, food such as meat and vegetables should be prepared in different utensils to avoid cross-contamination. Only 3.7% in preparing foods from home and transported the ready-to-eat foods to the market. Pre-preparation refers to a stage during which street food vendors purchase food, and some activities are conducted at home. Food is transported to the site for the final stage of cooking or heating before serving.

In our previous work [26], we reported that vendors (93.3%) prepared food at the vending site. Three-quarters of the owners (74.5%) were prominent people cooking food, and very few were assisted by the employees (23.9%) or spouses (1.6%). Processing activities can be an important source of jobs and income for women. Hence, it is essential to note the street food enterprise's role. On average, SFV could at least be able to pay R1400 employees monthly. At the same time, others reported spending more money on things such as electricity (R200) and gas (R430) weekly to produce food.

Predominantly cooked food was meat (98.8%) and Millie pap (stiff porridge) (97.1%), a South African staple food. Three quarters (75.3%) of the vendor's cooked vegetables and 34% were served as gravy and chakalaka (8.2%). The various meats cooked by the vendors were chicken (88.8%), beef (68.3%), beef tripe/*magulu* (14.3%), and beef sausage (5.5%), with very few street vendors selling fish (1.4%) or pork (2.2%). The median daily cost of meat was R400. The street food vendors cooked and sold two main starchy foods, mielie pap (97.1%) and rice (14.1%), at R70 daily.

3.3.1.6 Consumption/point of sale

Table 3 shows the frequencies and their associated percentages relating to the street vending target market and mode of marketing. The main street food consumers were mainly government officials and school children attending nearby schools. Rural customers who visit the town and spend some time in town also buy from street food vendors. Foods were either served at the stall or sold as takeaways. The study assessed the mechanisms street food vendors employ to sell their street foods. The results showed that 51 percent of the respondents sold their products to rural customers (82.2%) and school children (41.5%). Some street vendors reported their customers as government employees (59.9%) and middlemen (53.0%). The technologies used to market were call/WhatsApp (62.4%). Most street food vendors reported relying on face-to-face visits with customers to sell their foods. This method proved effective as per the saying "the proof of the pudding is in the eating". Personal visits validate the taste, flavours, the look and the eating experience that customers prefer.

Variables	Frequency*	Percentage (%)
Target market		
Rural customers	415	82.2%
School children	209	41.5%
Government employees	300	59.5%
Middlemen	267	53.0%
Other	9	1.8%
Mode of marketing		
Personal visits	368	72.0%
Phone call/WhatsApp	319	62.4%
Recruit professionally	22	4.3%
Word of mouth	14	2.7%
Posters	3	0.6%
Nothing	7	1.5%

**Number of responses.*

Table 3.
Frequencies and associated percentages street food vending target market and mode of marketing.

3.4 Value chain stakeholders and their role along the value chain

Table 4 present the major stakeholders and their specific roles. Firstly, we identified critical functions. Secondly, the actors involved then their roles were also outlined. As highlighted earlier, one of the critical functions during purchasing was providing supplier input to the vendors. In this study, various actors identified as inputs suppliers (Supermarkets, farmers, butcheries, hawkers) were the most dominant among the identified inputs suppliers. The second important function noted was processing where street food vendors were the actors in the chain activities involved in processing foods either at home or at the stall. The other stakeholders identified in the study were the regulations and quality assurance role players such as the local Municipality and the Department of Health. Lastly, customers are also identified as role players because they are the consumers of the final product of the street food.

3.5 Business enabling environment

3.5.1 Institutional supports

Table 5 presents the types of institutions available along the street vendor's chain. To improve the competitiveness of the street food value chain, every actor or stakeholder has a vital role to play. The current study explored the institutions available

Step I	Step II	Step III	
Critical function	Current/ potential actor	Specify role	
Inputs suppliers (purchasing)	Main supermarket	Spar	Supply raw and processed foods
		Boxer	
		Shoprite	
		OBC	
	Farmers		Vegetables and meat suppliers (pork and broilers)
	Butchers		Supply all types of meat
	Street vendors/ hawkers		Broiler and vegetable suppliers
Processing	Street food vendor		Production, cleaning, marketing, and selling
Consumption	Customers		Buy products
Regulations and quality assurance management	Local government or municipality Department of health		Provision of the following services: • Area management and placement • Collect waste/clean and issuing of permits/ licenses • Policy amendments
Other stakeholders	Media		Advertise business on social media platforms such as WhatsApp and Facebook

Table 4.
Major stakeholders involved in the street food and their roles.

Institutions	*N*	%
Formal lenders (banks)	4	1
Informal lenders	5	1
NGO	8	2
Associations	45	9
Local municipality	202	38
Department of health	235	46
Inputs suppliers	17	3

Table 5.
Frequencies and associated percentages on the types of institutions available along the street vendors' value chain.

that provided support services to ascertain which institutions were present and accessible amongst the street food vendors to provide various forms of support to the value chain actors. There is a relationship amongst the value chain actors, which was established based on spot markets (actors negotiate on price, quantities, and other requirements directly at the market).

3.5.1.1 Financial institution

The current study highlighted a poorly developed street food enterprise financial system. Although formal and informal institutions were available to assist street food vendors financially. Street food vendors could only get assistance in the form of loans from money lenders (1.0%) and very few banks (0.8%). Input suppliers (street food vendors/bakkie sellers) (3.3%) could also provide street food vendors with credits for a later return. Those who borrowed money from the informal money lenders (loan) sharks highlighted conditions including paying money later with interest, while some indicated paying back the money the same day without interest.

3.5.2 Regulations and quality assurance management

3.5.2.1 Street food vending associations

About nine percent of the street food vendors were members of associations or unions. Most of the representatives were from the street food vendor's associations at a local level. The in-depth interviews explored why street food vendors have not been affiliated or part of the street food association. Among the identified reasons, street food vendors were unaware of any association (6.9%). In contrast, others had no reason why they were not part of the association (27.5%). Street vendors who were not members of any representations provided the following reasons: "*I am a foreigner*," while others indicated that "*Associations only represent taxi drivers.*" Others highlighted that they were not aware of available associations of street food vendors. At the same time, most (64.1%) did not have reasons why they were not part of an association. The low participation rate of SFVs in associations is because these latter ones often lack clear organizational goals.

Regarding the street food vendors who were members of an association (8.5%), most street food vendors could not tell the role played by the associations, while some indicated that associations provide them with business advice and assist with solving problems. Street food vendors indicated that the street food vending association needs assistance to improve service delivery for their businesses to be competitive. Amongst the recommendations, the street food vendors include training, assistance in building better markets, and provision of adequate space for business operations. The street food vendors recommended access to quality infrastructure as the primary service improvement intervention for improving street food enterprise.

3.5.2.2 Local government

Based on the in-depth interviews conducted regarding the roles of the available stakeholders. Street food vendors highlighted that the main role of the local government (Municipalities) (38.3%) was to provide services such as allocation of placement or vending sites and management of vending spots. Local municipalities' role also included issuing permits or selling licenses at an allocated premise. However, not all street food vendors had licenses, as others had no interaction with officials except the suppliers and the customers. The local Municipality's role was to clean and collect waste daily. Other stakeholders who played a role in the street food sometimes the Department of Health (46.0%). They provided certificates of acceptance to sell cooked foods to street food vendors. However, not all vendors reported interacting with them to obtain the acceptance certificate to sell street food. Few vendors noted that NGOs indicated that they sometimes provide training from the NGOs.

3.6 Distribution of profit margins along with inputs to suppliers of street food vendors

3.6.1 Cost, output, and revenue

Table 6 shows the estimated profit margins and street food vendors' inputs suppliers.

3.6.1.1 Computation of cost, output, and revenue

We computed profit margin and ratio to determine if street food enterprise production was profitable in the study area. Four different types of plate production (chicken and beef meat) were used to compute the cost of production. Profit margin (*PM*) was calculated as the difference between the cost of production (*R*) and selling price OR revenue (*R*) (expressed as $P = R - C$). The percentage profit margin (*G*) was computed as the profit (*P*) divided by selling price OR revenue (*R*) (expressed as net sale-cost of goods)/net sale. Generally, street foods are sourced from the supermarkets and informal markets such as bakkies sellers/street vendors and farmers. The description of the activities done by the value chain actors, from the suppliers of the inputs to consumption (sales), was used to estimate the variable costs and returns.

Computations were performed per type of plate sold from different street food vendors daily. A typical plate sold by the street food vendor consists of meat, pap, green leafy vegetables served with a stew of tomato/onion. We used the two most cooked meat in this study to determine the profit margin, as shown in **Table 6**—the distribution used for daily production of chicken and beef plate profit margins from supermarkets and informal traders.

Product inputs suppliers	Production and marketing cost	*R*/unit	Total (*R*)	*Profit margin
Scenario 1: Inputs from formal traders (supermarkets/butcheries) with no labour required	Inputs production variable cost (veg, beef and maize meal) (IPVC)	49 + 279 + 70	398	78%
	Expenses (transport) (*E*)	11.3	11.3	
	Sales (selling price × total produced) (*S*)	43 × 44	1892	
	NI = *S* – *E*	1892 – (398 + 11.3)	1482.7	
	Profit margin = NI/*S*	1482.7/1892	0.78	
Scenario 2: Inputs from formal traders (supermarkets/butcheries) with labour	Inputs production variable cost (veg, beef and maize meal) (IPVC)	49 + 279 + 70	398	76%
	Expenses (transport, labour/day)	11.3 + 47	58.3	
	Sales (selling price × total produced) (*S*)	43 × 44	1892	
	NI = *S* – *E*	1892 – (398 + 58.3)	1435.7	
	Profit margin = NI/*S*	1435.7/1892	0.76	
Scenario 3: Inputs from informal traders (bakkies sellers/street vendors/ farmers) with no labour required	Inputs production variable cost (veg, live chicken and maize meal) (IPVC)	21 + 180.83 + 70	271.83	76%
	Expenses (*E*) (transport)	11.3	11.3	
	Sales (*S*) (selling price × total produced)	40 × 30	1200	
	NI = *S* – *E*	1200 (271.83 + 11.3)	916.87	
	Profit margin = NI/*S*	916.87/1200	0.76	
Scenario 4: Inputs from informal traders (bakkies sellers/street vendors/ farmers) with labour	Inputs production variable cost (veg, live chicken and maize meal) (IPVC)	21 + 180.83 + 70	271.83	72%
	Expenses (transport, labour/day)	11.3 + 47	58.3	
	Sales (selling price x total produced)	40 × 30	1200	
	NI = *S* – *E*	1200 – (271.83 + 58.3)	869.87	
	Profit margin (PM) = NI/*S*	869.87/1200	0.72	

**Profit margin: sale – expenses/sales × 100*
Informal input suppliers are when street food vendors buy their inputs from local farmers/street vendors/bakkie sellers.

Table 6.
Estimates of street vending profit margins.

To determine the street food vendors' profit margin, we used the formula: Formula; profit = price of plate × total plate sold/expenditure. Therefore, it was calculated as follows, profit margin (PM) = net income (NI)/sales.

a. Where NI = sales (total plate sold × price of a plate) – expenses (inputs cost from supermarkets + cost of transport + labour).

b. Formal markets referred to street food vendors purchasing inputs from the supermarket.

3.6.1.2 Profit margins in the four scenario mapped

Four scenarios/channels were used to estimate profit margin from purchasing inputs from the formal and informal markets cost. Considering the difference between the price of chicken and beef sold in the informal markets and supermarkets, all street food vendors made a profit even though the profit margins differed.

a. **Scenario 1**: The findings of marketing Scenario 1 highlight that when purchasing inputs from supermarkets and no labour is required for the production, street food vendors could make 78.0% of profit on a beef based plate considering the expenses and sales of the day.

b. **Scenario 2**: If Street food vendors purchase their inputs from the supermarket and require labour for daily production, the vendor's profit margin could be 76.0%, for beef-based plate considering the expenses and sales of the day street food vendors. This highlights that even if street food vendors purchase their inputs from supermarkets, they do make enough profit with or without the extra service labour.

c. **In Scenarios 3 and 4**, street food vendors purchased inputs from informal traders. The difference was that in Scenario 4, street vendors have at least one assistant paid. However, the two scenarios highlighted that street food vendors purchasing their inputs for daily production could still make enough profit, on a chicken-based plate which was 76% for scenario 3 and 72% for scenario 4, respectively.

4. Discussions

4.1 Structure of the current street food value chain in the Vhembe district

This study aims to analyze the value chain of foods sold by street food vendors in the Vhembe district town of the Limpopo province. This was done by identifying the prominent value chain actors, institutions governing the chain, and key factors and challenges affecting the success or failure of the value chains for street foods. The street food value chain, faces several constraints [24, 32, 33], and this study confirms such findings. The value chain of food sold on the street is short, with no infrastructure. Complete value chain actors are linked in the chain processes necessary for transforming and transporting raw materials from suppliers to consumers. The activities during this period included purchasing, transportation, storage, production, and consumption. These activities of the food production chain in the current study were similar to a study by Cortese et al. [16], where they highlighted that in street food enterprises, the chain of activities starts from the acquisition of raw materials to service to the consumer a step which was earlier proposed by Barro et al. [34], as **Figure 1** illustrates.

4.2 Market channel

The first marketing channel was from the informal traders (i.e., street vendors/hawkers) to street food vendors traders. The other marketing channel was formal traders (i.e., supermarkets/wholesalers) and street food vendors. The end market of street food is domestic consumption.

Like other studies in South Africa and developing countries, street food vendors purchase their inputs from other vendors. Adeosun, Greene, and Oosterveer [35] also noted that most vendors stock their raw food materials to prevent having to go to the market daily. Street food vendors also buy copious quantities to prepare for a day until their stock get finished as a coping mechanism to deal with waste and food safety. This was corroborated by other studies conducted in India [36, 37]; Ghana [38] and Sabbithi *et al.* [39] in India. It was also established that Street Food Vendors relied on local food value networks with both the formal and informal upstream value chain [40]. A dualist position was noted in the current study whereby street food vendors' inputs were sourced from direct sources such as bakkie/street vendors and the local supermarket.

4.3 Value chain activities

4.3.1 Value chain actors

VCA of street food enterprise in the Vhembe district is simple and undeveloped with little infrastructure. The main actors in the value chain were input suppliers (smallholder farmers, traders/retailers, street vendors/hawkers) and consumers. The current study showed that while street food enterprise is embedded into informality, selling cooked foods in the rural towns highly relies on inputs from formal sector enterprises in the form of agricultural producers, wholesalers, and retailers (**Figure 4**).

4.3.1.1 Inputs suppliers

Among the current study's chain actors were the inputs' suppliers. Input suppliers bring together a range of products and present them in a way that is convenient to customers. According to Porter [19], these are related and supporting industries. Suitable suppliers of inputs are crucial for sustainable production and where traceability, environmental concerns, and quality assurance are involved [41]. Moreover, in other developing countries, there is a dualistic relationship between the formal and informal food sectors in exchange for inputs supplies [42]. Street food vendors currently have a direct relationship with supermarket because food cooked by street food vendors are commonly sold in the supermarket, indicating guaranteed and increasing demand for street food in the Limpopo Province. Tawodzera and Crush [43] study found that location informed purchase for 49%, followed by price (42%), and quality (9%). However, Senyolo *et al.* [44] some informal traders in the Limpopo province have established relations to some supermarkets (such as SPAR, Boxer, Pick and Pay and OBC, Shoprite) selling cooked food mainly by the street food vendors, including the green leafy vegetables. These findings collaborate with other studies in the country [45].

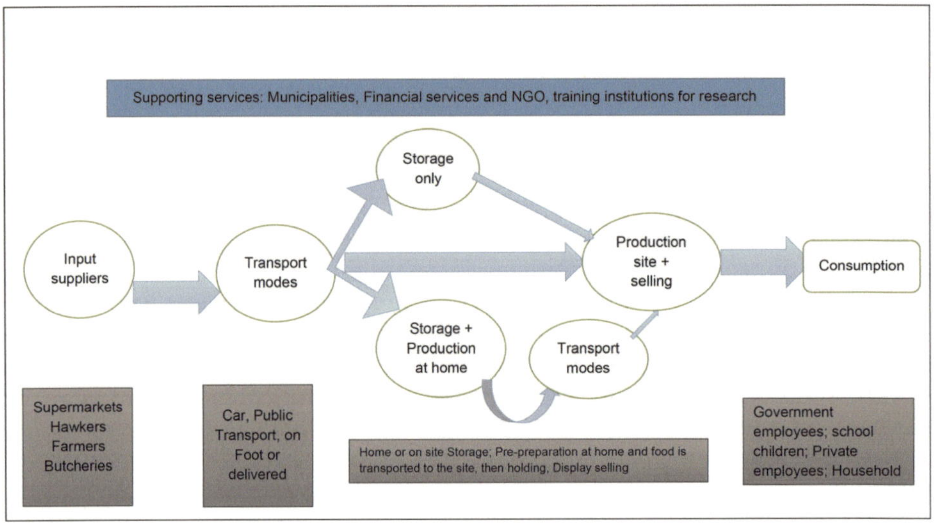

Figure 4.
Presentation of identified street food chain activities.

Moreover, Tawodzera and Crush [43] noted 22% of households rely on the agricultural production of their products. Contrary to the current study where very few street food vendors highlighted using their own produce for the vending instead relied on others for their enterprises.

The informal traders in the study supplied green leafy vegetables and live chickens to the street food vendors at a reasonable price to the formal traders for better profit margins.

4.3.1.2 Consumers

According to a survey by FAO [46], approximately 2.5 million people across the globe consume street foods every day. Most consumers are between the ages of 26–35 years, out of which most are single males away from home. Moreover, other essential categories of consumers include children, students, and office workers [11]. Charman *et al.* [7] also noted that residential micro-enterprises such as street vendors serve immediate resident consumer demands, primarily for essential fast food. Like the current study, street food enterprises' main customers mainly were people who travel to the rural town searching for work and school children during their lunch breaks. This study also noted that street food vendors generated demand for services provided by formal sector public and private actors, including transportation and formal shops, as reported by Roever and Skinner [47]. In Cape Town Hill *et al. [48]*, most street food consumers are single black males with some high-school education and matriculated. Moreover, the patterns of consumption of street food daily and two to three times a week by 38% and 43.3%, respectively.

4.4 Stakeholders' relationship in the street food chain (institutional support)

Thuy *et al.* [49] stated that the availability of various stakeholders in a value chain, such as intermediaries and traders, adds value to the value chain of a product. As the small scale of value chain linkages is prominent in the cities. The current study highlighted a relationship between the suppliers of the formal and informal inputs. However, street food vendors were the primary buyer of inputs from these different suppliers. The formal and informal street food input suppliers could establish a relationship that would benefit the street food enterprise by stabilizing input prices [44]. Thus, contributing to the distribution of the product in a geographic term.

4.4.1 Financial institutions

Food vendors also introduced some innovations to improve the food supply in the urban area. Haleegoah *et al.* [24] indicated that if street food vendors are given an opportunity for financial assistance, there could come up with innovative improvements in the daily operations of their enterprises. In the current study, street food vendors were limited due to a lack of financial support from recognised institutions such as banks and the government. Osei-Mensah *et al.* [32] found that inadequate managerial skills and financial constraints negatively affected the gross margin ratio. This shows the extent to which a lack of support and cooperatives is given to street food vendors in the Vhembe district.

Government legislation, regulations and policies can constrain value chain upgrading, amongst other ways, by setting trade barriers for production materials and production technology, limiting the flow of information, national and international, by imposing unfavourable taxes, and denying infrastructural investments that would benefit value chains.

Even though street food vendors noted concerns that inputs sold from the formal traders are expensive compared to the informal traders. In the current study, we observed poor institutional support for street food vendors along the value chain of street food. In previous publication [26, 27], we highlight that street food vendors have agreements amongst those selling vegetables whereby they take inputs on credits to pay after the sale, something absent from the formal retailer as street food vendors are not allowed to make any credits from the formal institutions. The study highlighted the support of the informal sectors such as transport because most street food vendors used public transport to fare the input to their stalls [40].

4.4.2 Regulatory and quality assurance institution

4.4.2.1 Street food vending organizations

Inability to sustain associations had been identified as a challenge to food vending [24], which our findings corroborate. This has prompted the establishment of innovation platforms in the street food enterprise for which studies and other training could be carried out. **Table 4** reports the roles and functions of the available affiliated associations. However, in-depth interviews explored why street food vendors have not been affiliated or part of the street food association. Among the identified reasons, street food vendors were unaware of any association (6.9%). Studies have shown that only a few vendors were known to be part of an association [24, 32]. Moreover, Wills [28] states that the absence of organisational goals could be the main reason for these associations' not thriving among food vendors.

As a first step towards recognition, street food vendors should constitute themselves into -organized associations that would enable them to develop a code of practice for their businesses or constitute a forum for interaction with the relevant authorities.

4.4.2.2 Local government

The most common problem with street vending is issues with local authorities [24, 32, 33, 50, 51]. In the current study, we observed non-compliance by street food vendors to obtain necessary documentation before operation. While on the other site, street food vendors highlighted that there is neither support nor interaction between street food vendors and the local authorities to ensure proper street food value operations.

A collaboration between the street food vendors and the relevant stakeholders, such as the local Municipality, should be strengthened. Haleegoah *et al.* [24] also indicated that this poor interaction between the authorities and the street food vendors creates conflict among the stakeholders; thus, city planners should designate appropriate areas within the towns for local food vending could enhance compliance with safe and environmental regulations.

4.5 Regulations and quality assurance management

Although Roever and Skinner [47] indicated the importance of policies, laws, standards, regulations, and institutional support services forming the chain environment in the street food value chain for better performance, street food vendors do not consider obtaining specific requirements, including certification before vending and obtaining a certificate of acceptance for selling foods to the public (REF-SA). This practice is common in other African countries, such as Ghana, a street food vendor requires medical screening and certification before vending [38]. However, this health demand seems unsuccessful as it is noted in Ghana and South Africa that street food vendors are unaware of their requirements before establishing an enterprise.

Like the current study, only 38.3% and 46.0 of the street food vendors were aware that they needed to engage with the local Municipality and the department of health to obtain a license to operate as street food vendors in the Vhembe district. In line with this, other studies [52] have shared similar views, where most of the food vendors interviewed had no health certificates but operated as vendors. Therefore, vendors must adhere to high hygiene standards when dealing with food.

4.6 Profit margin and value shares

The current study highlighted that street food vendors purchase inputs from formal and informal traders. A higher profit margin was obtained in purchasing goods from the formal retailers on a beef-based plate at 78% profit margin and 76% with extra labour. However, the profit margin percentage, made at the current production cost purchases from the informal traders was 76% profit margin and 72% with extra labour. According to Poojara and Dhanesh [11], street food vendors have a minimal profit margin and are incentivized to keep expenses low by utilizing low-quality ingredients and disregarding costly hygienic practices. Factors such as the cost of labor and the other expenses contributed to the distribution of different profit margins of various plates sold. Vendors with no added costs, such as labor costs, make more profit than those with labor. Similar trends were also observed by Sugri et al. [53], where it was indicated that additional attributes, and climatic conditions could increase the input cost from the suppliers and those purchase might have a lower return.

5. Conclusion and recommendation

The profit margin percentage analyses revealed that street food vending enterprise productions were profitable. The profit margins were 72–78% depending on the costliest meat product in the plate. Beef-based scenarios seem to fetch better profit margins (76–78%) based on the higher prices and increased number of plates sold. The chicken-based profit margins were at the range of 72–76%. The price per plate were lower with a smaller number of plates sold. There should be improved linkages between the local authorities, street food organizations, formal (formal and informal), and financial institutions to the street food vending enterprise with a collective profit motive.

In addition, measures should be put in place to solve the street food vending constraint affecting their operation, such as lack of access to finance, poor infrastructure and lack of managerial skills, and the cost of inputs. If street food vending enterprises could be organized aligned with roles of various actors, this could minimize the risk of failure of the street food enterprises sector. Due to the high demand for street food and its affordability in rural towns, authorities such as the department of health should assist street food vendors to minimize the risk of food safety along the chain. For rural and peri-urban communities street food vending could be a vehicle for improving the population's nutritional status and reduce hunger.

Acknowledgements

The authors are grateful to the participating street food vendors that provided information during this study. The authors acknowledge the field workers and the academic support from the universities (the University of the Free State and the University of Venda).

Conflict of interest

The authors declare that there are no conflicts of interest in the publication of this research paper.

Disclosure

The authors take full responsibility for any errors.

Abbreviations

NGO	Non-governmental organisations
SFV	Street food vendors
SPSS	Statistics package for social sciences
VCA	Value chain analysis

VC	Value chain
SWOT	Strength weaknesses opportunities threats

Author details

Tjale Clopus Mahopo[1,2], Noxolo Cebisa Nesamvuni[1], Azwihangwisi E. Nesamvuni[2]* and Johan van Niekerk[2]

1 Department of Nutrition, University of Venda, Thohoyandou, South Africa

2 Sustainable Food Systems and Development, Centre for Sustainable Agriculture, Rural Development and Extension, University of the Free State, Bloemfontein, South Africa

*Address all correspondence to: nesamvunie@gmail.com

References

[1] Draper A. Street foods in developing countries: the potential for micronutrient fortification. London School of Hygiene and Tropical Medicine. 1996:1-74

[2] Martins JH. Socio-economic and hygiene features of street food vending in Gauteng. South African Journal of Clinical Nutrition. 2006;**19**(1):18-25. DOI: 10.1080/16070658.2006.11734085

[3] von Holy A, Makhoane FM. Improving street food vending in South Africa: achievements and lessons learned. International Journal of Food Microbiology. 2006b;**111**(2):89-92

[4] World Health Assembly. The World Health Report 1996: Fighting Disease, Fostering Development. World Health Organization, 49. 1996. Available at: https://apps.who.int/iris/handle/10665/178481

[5] Skinner C, Haysom G. The informal sector's role in food security: A missing link in policy debates? Working Paper 44. Cape Town: PLAAS, UWC and Centre of Excellence on Food Security. 2016;**12**:II–16

[6] Abrahale K, Sousa S, Albuquerque G, Padrão P, Lunet N. Street food research worldwide: a scoping review. Journal of Human Nutrition and Dietetics. 2019;**32**(2):152-174. DOI: 10.1111/jhn.12604

[7] Charman AJE, Petersen LM, Piper LE, Liedeman R, Legg T. Small area census approach to measure the township informal economy in South Africa. Journal of Mixed Methods Research. 2017;**11**(1):36-58. DOI: 10.1177/1558689815572024

[8] Maloney WF. Informality revisited. World Development. 2004;**32**(7):1159-1178

[9] Cohen MJ, Garrett JL. The food price crisis and urban food (in)security. Environment and Urbanization. 2010;**22**(2):467-482. DOI: 10.1177/0956247810380375

[10] Even-Zahav E, Kelly C. Systematic review of the literature on 'informal economy' and 'food security': South Africa, 2009–2014, Working Paper 35. Cape Town: PLAAS, UWC and Centre of Excellence on Food Security; 2016

[11] Poojara, R and Dhanesh, B. 2021. Street Foods: Safety Concerns. New India Publishing Agency, Pitam Pura, New Delhi, India. Web: www.nipabooks.com

[12] Jordaan H, Grové B, Backeberg GR. Conceptual framework for value chain analysis for poverty alleviation among smallholder farmers. Agrekon. 2014;**53**(1):1-25. DOI: 10.1080/03031853.2014.887903

[13] Laforteza JJ, Manalo G, Joel T. Determinants of accounting practices among street food vendors. International Journal of Accounting and Financial Management Research. 2014;**4**(3):2249-6882

[14] Kaplinsky R, Morris M, Readman J. Understanding upgrading using value chain analysis. Bam. 2002;**2002**:1-13

[15] Bwalya R, Kalinda T. An analysis of the value chain for indigenous chickens in Zambia's Lusaka and Central Provinces. Journal of Agricultural Studies. 2014;**2**(2):32. DOI: 10.5296/jas.v2i2.5918

[16] Cortese RDM, Veiros MB, Feldman C, Cavalli SB. Food safety and hygiene practices of vendors during the chain of

street food production in Florianopolis, Brazil: a cross-sectional study. Food Control. 2016;**62**:178-186. DOI: 10.1016/j.foodcont.2015.10.027

[17] Donovan J, Franzel S, Cunha M, Gyau A, Mithöfer D. Guides for value chain development: a comparative review. Journal of Agribusiness in Developing and Emerging Economies. 2015;**5**(1):2044-0839. DOI: 10.1108/JADEE-07-2013-0025

[18] Barr A. 'An Analysis of the Factors Impacting the Competitive Performance of the South African Wine Industry Value Chain', no. April: 1–133. 2019 Available at: http://hdl.handle.net/10019.1/105864

[19] Porter ME. The Competitive Advantage of Nations. London: McMillan; 1998

[20] Mathaulula MA. Pillars of a safety management system for small-scale vended foods in a rural-based municipality of South Africa. 2020

[21] Mchiza Z, Hill J, Steyn N. Foods currently sold by street food vendors in the Western Cape, South Africa, do not foster good health. In Fast Foods: Consumption Patterns, Role of Globalization and Health Effects Journal of Hunger & Environmental Nutrition 2014a;**14**(3):1-15. DOI:10.1080/19320248.2018.1434104

[22] Mathaulula MA, Francis F, Mwale M. Contribution of small-scale food vending to rural livelihoods in Thulamela Municipality of Limpopo Province, South Africa. Studies of Tribes and Tribals. 2015;**13**(1):40-47. DOI: 10.1080/0972639x.2015.11886710

[23] Nyahunda L, Tirivangasi HM. Effects of climate change on rural women in Makhado Municipality, Vhembe District, Limpopo Province, South Africa. Gender and Behaviour. 2020;**18**(2):15769-15778

[24] Sarpong HJA, Akuoko KO, Dwumah P, Marfo AF, Forkuor JB, Frimpong BN, et al. Actors perceptions on challenges and opportunities to improve street vended local foods in Urban Ghana. Advances in Applied Sociology. 2020;**10**(11):435-453. DOI: 10.4236/aasoci.2020.1011025

[25] Chauke PK, Munzhelele R, Maiwashe A. Some factors impacting on street sellers' ability to generate above poverty line income in Vhembe District, South Africa: logistic regression approach. Journal of Social Sciences. 2015;**44**(1):8-14. DOI: 10.1080/09718923.2015.11893452

[26] Mahopo TC, Nesamvuni CN, Nesamvuni AE, De Bruyn M, Van Niekerk J. Socioeconomic characteristic of street food vending enterprises in the Vhembe District, Limpopo Province. Technium Social Sciences Journal. 2022a; **29**(2012):419-437. DOI: 10.47577/tssj.v29i1.5944

[27] Mahopo TC, Nesamvuni CN, Nesamvuni AE, de Bryun M, van Niekerk J, Ambikapathi R. Operational Characteristics of Women Street Food Vendors in Rural South Africa. Frontiers in Public Health. 2022;**10**:849059. DOI: 10.3389/fpubh.2022.849059

[28] Wills G. Women in informal employment: globalizing and organizing South Africa's informal economy: a statistical profile. WIEGO Working Paper (Urban Policies). doi:10.1016/j.wasman.2013.10.045

[29] Veal AJ. Research Methods for Leisure and Tourism: A Practical Guide. 4th ed. Harlow: Pearson Education Limited; 2011

[30] Alvi MA. Manual for Selecting Sampling Techniques in Research. Mohsin Hassan Alvi: University of Karachi, Iqra University; 2016

[31] Yamane T. Statistics: An Introductory Analysis. 2nd ed. New York: Harper and Row; 1967

[32] Osei-Mensah J, Ohene-Yankyera K, Aidoo R. Constraints to growth of micro and small-scale enterprises in Ghana: a case of street food enterprises. Journal of Development and Agricultural Economics. 2016;**8**(10):241-250. DOI: 10.5897/jdae2016-0732

[33] Resnick D, Sivasubramanian B, Idiong IC, Ojo MA, Tanko L. The enabling environment for informal food traders in Nigeria's secondary cities. Urban Forum. 2019;**30**(4):385-405. DOI: 10.1007/s12132-019-09371-7

[34] Nikiema P, Barro N, Razack B, Itsiembou Y, Savadogo A, Tidiane OA, et al. Street-Vended Foods Improvement: Contamination Mechanisms and Application of Food Safety Objective Strategy: Critical Review. Pakistan Journal of Nutrition. 2007;**6**(1). DOI: 10.3923/pjn.2007.1.10

[35] Adeosun KP, Greene M, Oosterveer P. Informal ready-to-eat food vending: a social practice perspective on urban food provisioning in Nigeria. Food Security. 2022;**763–80**. DOI: 10.1007/s12571-022-01257-0

[36] Choudhury M, Mahanta L, Goswami J, Mazumder M, Pegoo B. Socio-economic profile and food safety knowledge and practice of street food vendors in the City of Guwahati, Assam, India. Food Control. 2011;**22**(2):196-203. DOI: 10.1016/j.foodcont.2010.06.020

[37] Reddy AA, Ricart S, Cadman T. Driving factors of food safety standards in india: learning from street-food vendors' behaviour and attitude. Food Security. 2020;**12**(6):1201-1217. DOI: 10.1007/s12571-020-01048-5

[38] Dun-dery EJ, Addo HO. Food hygiene awareness, processing and practice among street food vendors in Ghana. Journal of Food and Public Health. 2016;**6**(3):65-74. DOI: 10.5923/j.fph.20160603.02

[39] Sabbithi A, Naveen Kumar R, Kashinath L, Bhaskar V, Sudershan Rao V. Microbiological Quality of Salads Served along with Street Foods of Hyderabad, India. International Journal of Microbiology. 2014:1-6. DOI: 10.1155/2014/932191. Epub 2014 May 14. PMID: 24955092; PMCID: PMC4052067

[40] Petersen LM, Charman AJE, Kroll FJ. Trade dynamics in cape town township informal foodservice–a qualitative and supply chain study. Development Southern Africa. 2018;**35**(1):70-89. DOI: 10.1080/0376835X.2017.1412297

[41] Department of Agriculture, Forestry and Fisheries (DAFF). Biodiversity Risk and Benefit Assessment (BRBA) of Alien Species in Aquaculture in South Africa. South African Molluscan Shellfish Monitoring & Control Programme Annual Report. 2011

[42] Bouafou GM, Beugré KGFC, Amani YC, Bouafou KGM. Street food around the world: a review of the literature. Journal of Service Science and Management. 2021;**14**:557-575. DOI: 10.4236/jssm.2021.146035

[43] Tawodzera G. Food vending and the informal sector in Cape Town, South Africa. Hungry Cities Partnership. 2019;**23**:1-14

[44] Senyolo GM, Wale E, Ortmann GF. Analysing the value chain for african

leafy vegetables in Limpopo Province, South Africa. Cogent Social Sciences. 2018;**4**(1):1-16. DOI: 10.1080/23311886.2018.1509417

[45] Skinner C, Haysom G. Informal food retail and food security in Africa: Myths and facts. Consuming Urban Poverty Policy Brief. 2018;**1**

[46] FAO. The state of food insecurity in the world. Rome; 2001

[47] Roever S, Skinner C. Street vendors and cities. Environment and Urbanization. 2016;**28**(2):359-374. DOI: 10.1177/0956247816653898

[48] Hill J, Mchiza Z, Fourie J, Puoane T, Steyn N. Consumption Patterns of Street Food Consumers in Cape Town. Journal of Family Ecology and Consumer Sciences = Tydskrif Vir Gesinsekologie En Verbruikerswetenskappe. 2016, 2016; (Special Edition 2):25-35

[49] Thuy NT, Man PD, Lebailly P, Le Thi C, Minh HN, Thu T, et al. Mapping the pork value chain in vietnam: a systematic review. Tropical Animal Health and Production. 2020;**52**(6): 2799-2808. DOI: 10.1007/S11250-020-02338-Y

[50] Edeme RK, Nkalu NC. Operations of street food vendors and their impact on sustainable life in Rural Nigeria. American Economic & Social Review. 2018;**4**(1):1-7. DOI: 10.46281/aesr.v4i1.207

[51] Tshikhudo LP, Manenzhe TD. Evaluation of the profitability of vegetable and fruit street vendors in Mbombela local municipality in Mpumalanga Province, South Africa. European Journal of Agriculture and Food Sciences. 2021;**3**(2):49-52. DOI: 10.24018/ejfood.2021.3.2.227

[52] Nkosi NV, Tabit FT. The food safety knowledge of street food vendors and the sanitary conditions of their street food vending environment in the Zululand District, South Africa. Ultima Management : Jurnal Ilmu Manajemen. 2014;**7**(1):37-46. DOI: 10.1016/j.foodcont.2010.06.020

[53] Sugri I, Maalekuu BK, Gaveh E, Kusi F. Sweet potato value chain analysis reveals opportunities for increased income and food security in Northern Ghana. Advances in Agriculture. 2017; **2017**:1-14. DOI: 10.1155/2017/8767340

Chapter 6

Reducing Inequalities in the Coffee Value Chain: Threats and Opportunities for Small-Scale Farmers in Central America and East Africa

Ingrid Fromm

Abstract

The global trade of agricultural commodities has increased in the last decades, creating economic growth opportunities in many areas of the world. However, for smallholder farmers producing commodities traded in global markets, there is mixed evidence of a positive economic impact, particularly in coffee production. Rural incomes in Central America have remained stagnated and for smallholder coffee producers in East Africa, farm gate prices often do not cover production costs. Poverty and food insecurity are issues impeding development in the region. Yet empirical evidence points at global value chains as a driver for income growth and development in emerging and developing countries. As pointed out in The State of Agricultural Commodity Market 2020 global trade and inclusive, well-functioning markets with lower trade barriers, opportunities for technological upgrading, and access to markets can spur sustainable development. In this chapter, a comparative analysis of the inequalities and factors hindering the economic growth of coffee farmers in Central America and East Africa will be provided.

Keywords: sustainability, inequality, smallholder coffee farmers, Central America, East Africa

1. Introduction

Coffee is one of the world's most traded agricultural commodities. Both in terms of volume and value, it is listed as a top traded commodity, along with corn, wheat, soybeans, and sugar. The global coffee market was estimated to reach a value of US$ 114.1 billion in 2020 [1, 2]. Coffee is produced in the coffee-belt, a region 25–30° north and south of the Equator. About 40 countries in this tropical region grow most of the coffee traded in the world and estimates place 12.5 million farmers, predominantly smallholder farmers, as the main producers of coffee [3] In addition to production, a significant labor force is employed in the coffee sector worldwide. An estimated

125 million people worldwide work directly in the coffee value chain [4]. Most of the coffee production systems are small scale, and most of the coffee is cultivated in plots of land of 5 ha or less [3]. Thus, smallholder farmers are responsible for most of the global coffee production [4]. Large coffee estates of over 50 ha are rather the exception and are found mostly in Central and South America.

Coffee certainly is one of the world's most beloved beverages, and it is consumed across all geographic regions. Demand is also increasing, driving higher production. From 2008 to 2020, production increased from 8.5 million tonnes to 10.7 million tonnes of coffee beans [5]. Brazil, Vietnam, and Colombia lead the world production of coffee, accounting for 62% of the annual total production [6]. Other producing countries in the coffee bean belt include Central American countries like Honduras, Guatemala, Costa Rica, El Salvador, and Nicaragua. In East Africa, main producing countries include Kenya, Ethiopia, Tanzania, Uganda, and Rwanda. Both of these geographic areas in the bean belt are characterized by having ideal conditions for coffee cultivation, including high-altitude areas, temperatures somewhere between 21° and 29°C and good soils, Arabica coffee plants thrive in these conditions. For these producing countries, coffee is of major economic importance, as the sector is an important source of their export revenue, which can be significant. In Ethiopia, for example, coffee accounts for more than a quarter of the export earnings and in Burundi, coffee represents at least 20% of the national export revenue. In Uganda and Honduras, the coffee export revenue accounts for about 10% of the export earnings [7].

2. Global trends

Despite the economic importance of coffee, most small-scale farmers in Central America and East Africa face similar conditions of poverty. Covering production costs is challenging, and earning a decent income as coffee producers is difficult. The coffee value chain is complex, and multiple interactions take place across several geographical locations, so farmers have no decision power when it comes to issues such as the global market price of green coffee. Although there are some opportunities for economic growth, these are found in certain sectors like specialty, certified coffee but elude most conventional farmers. For the most part of history, coffee has been produced and traded as an undifferentiated commodity. Large companies which sold household coffee brands would buy bulk coffee, and there was little traceability, often not even the country of origin was known, as different lots coming from all over the world would be mixed. Thus, prices would vary according to species (Arabica or Robusta), but the origin was largely unknown by the end consumer.

Another reason for the disparities in the profit distribution have to do with the governance of the coffee value chain and the complexity of interactions, where numerous buyers and sellers transform coffee at different points and over various geographic areas. Farmers play a limited role in the transformation of coffee. They harvest the coffee, then bring it to a collecting and processing point. Not all farmers dry the coffee themselves. In Central America, these mills where coffee is first dried are known as "*beneficios*" or in East Africa, "wet mills." The main activity is to remove the pulp, wash, and dry the coffee. These facilities are usually located in convenient locations for further transformation, and that includes rivers and towns so the coffee can be transported easily from these points. Not all coffee growers are in the proximity of these stations, and some might face difficulties to bring the coffee to these mills. In Central America, some middlemen collect the coffee and bring the freshly picked coffee cherries to the "*beneficios.*" Once the coffee is received at the collection points or the *beneficios* or wet mills, farmers receive their payment for the coffee sold. Already at this point, traceability may be an issue, if there is no system in place to separate coffee by lots and location. Once transported to the *beneficio* and processed, the coffee beans from different regions are blended. In some cases, the mixing of lots might promote a uniform quality of coffee, combining the characteristics and flavor profiles of the different regions [8]. From there, the coffee is taken in charge by an exporter, who usually bring the coffee to another location to either further process the coffee or store it for export. In Central America, the export process is done through larger coffee importers or through the large coffee producer associations, which have established commercial links to large international coffee traders. These two parties take care of the transportation of the coffee from the *beneficio* or mills to the destination country, sometimes even to the buying roasteries. The roasting is usually done in the country of arrival, and the coffee is sold to local buyers, retailers, hotels, restaurants, or cafes, which handle of the distribution to the final consumers.

Global coffee consumption has steadily increased in the last decades, but there are variations across different regions. While coffee consumption has increased in Europe, North and South America, and Asia, it had remained stagnant in Central America and decreased in Africa. Industrialized counties experienced great changes in consumption patterns of coffee. Up until the 1980s, coffee was mainly consumed as a plain black cup of coffee, and there was little differentiation in origins, roasting or flavors. This has now changed and a multitude of forms, flavors, and origins, stemming from the coffee shop culture dramatically evolved since them. In parallel to the changes in consumption, ICO [9] reported an increase of 2% annually since 1990. Prices per cup of coffee have also increased, and consumers pay much more per cup of coffee than they did in 1990 [10]. Home consumption of coffee also evolved and so did the market, which now offers a wide range of coffee machines, brewers, presses, kettles, pots, and pour-over glassware which are much more sophisticated than the simple percolator commonly found in homes in the 1980s. The changes in coffee consumption patterns and the new trends for more differentiated coffee did not, however, lead to an increase of farm gate prices paid to the coffee farmers. Farmers in Central America have faced severe crises such as the coffee crisis of 2001 which severely impacted the region. East Africa experienced a similar stagnation during this time (**Figures 1** and **2**).

The consequences of low farm gate prices are a huge burden for farmers, particularly small-scale farmers, because as farm incomes decline, livelihoods are increasingly at risk. Such low incomes make it impossible for small-scale farmers to invest in the modernization of farms and even less in good agricultural practices which may lead to a more sustainable coffee production [11]. By and large, small-scale farmers in Central America and East Africa are finding it impossible to buy the very basic inputs to sustain the current coffee yields. In instances where inputs such as fertilizers are not subsidized, farmers have limited opportunities to pay for these inputs, especially with the current increased prices of fertilizers. In the last three decades, the average prices paid to the producers have fluctuated and in the 2010s, higher prices were paid to producers (**Figures 1** and **2**). However, prices have since dropped, making it difficult for the small-scale farmers in Central America and East Africa to make a profit in coffee production, especially since the 2021/2022 season, where the prices for fertilizers increased by up to 80% [12]. At the farm level, these increases in the prices of fertilizers mean that production costs will barely be covered if the coffee prices do not increase.

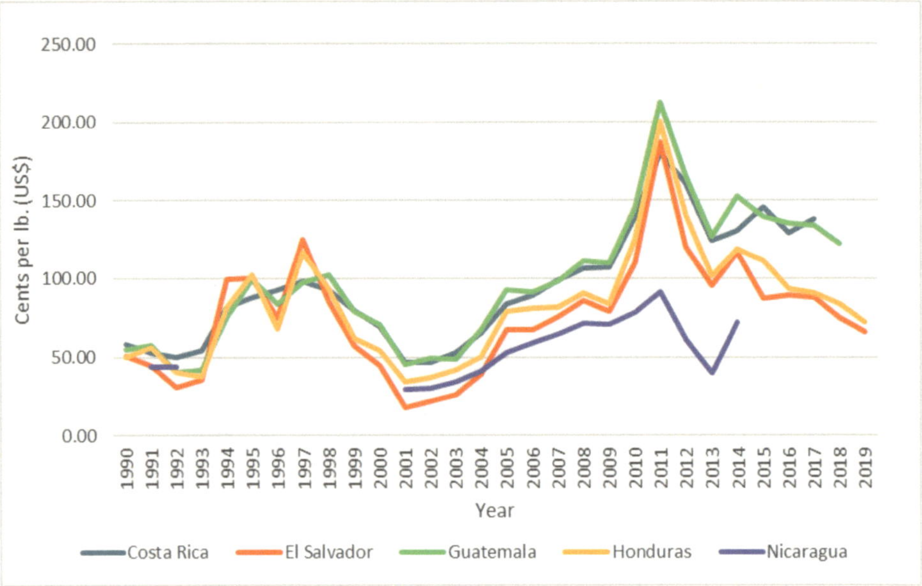

Figure 1.
Prices paid to producers in central American countries 1990–2019 (Source: [6]).

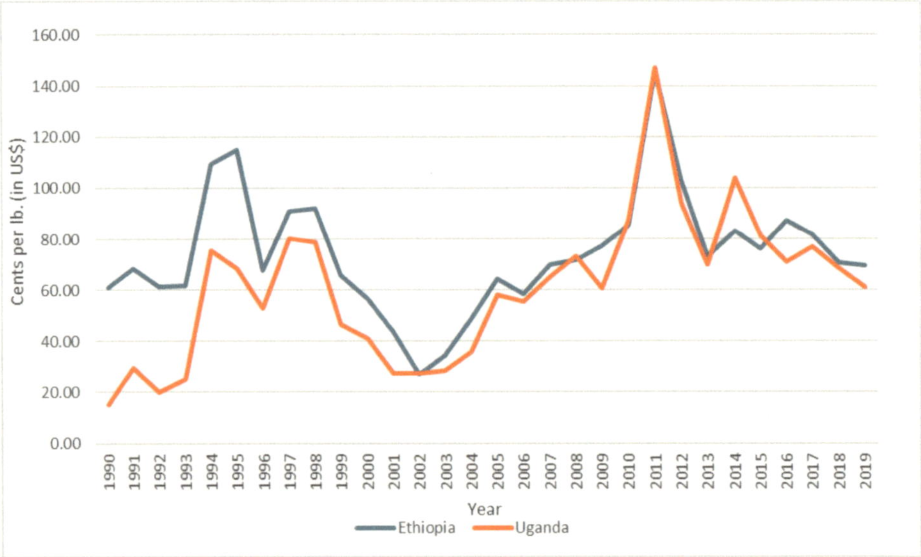

Figure 2.
Prices paid to producers in Ethiopia and Uganda 1990–2019 (Source: [6]).

3. The coffee value chain in Central America

Coffee is one of the top agricultural sectors in Central America, and in El Salvador, Honduras, and Nicaragua, it is the main agricultural export and second largest agricultural export in Guatemala and Costa Rica. In all countries, coffee is the main source of rural employment. An estimated 1.2 million people are directly employed in the coffee sector in Central America [9]. Although large estates, usually family owned, do exist in Central America, most of the production takes place in small-scale family farms. In rural areas where coffee is cultivated, poverty still prevails [13–15]. Farm gate prices are low and part of the problem of low income has to do with one main payment made to producers during the harvest season only. Fluctuating international market price patterns negatively impact farmers. In 2018, market prices dropped 30% below the average, which meant that farmers were not able to cover production costs. Such low incomes have long-lasting consequences which keep farmers poor. They are unable to make proper investments to boost productivity or implement sustainable production practices. Coffee farmers in this region are susceptible to climate change risks, which will likely impact their livelihood and in general, the future of coffee production [16].

In terms of climate change impact, the region, which in part lies in the dry corridor of Central America, is one of the most affected areas by climate change [17]. Rising global temperatures can affect coffee yields [9]. Higher temperatures are severe challenge for Arabica coffee production. To produce coffee, specific ecological and meteorological conditions are necessary. Temperature should ideally range between 15 and 23°C. In 2019, hydric stress severely affected fruit formation in coffee farms in some areas of Central America, thus lowering yields [18]. The already established high-altitude coffee areas will no longer find the ecological conditions needed to maintain that quality, thus pushing the agricultural frontier to higher-altitude regions. If coffee production expands to higher-altitude areas, protected forest areas will be affected. In the lowlands where coffee is also cultivated, higher temperatures and dry spells will likely result in lower yields.

In addition to increasing temperatures and droughts, sever climate events such as hurricanes have also significantly impacted smallholder coffee producers. Extreme impacts such as tropical storms and hurricanes with strong winds and torrential rainfall destroy coffee plantations. In November 2020, when the coffee harvest was about to start in the region, Hurricanes Eta and Iota, two category 5 and 4 hurricanes unprecedently hit the area in the lapse of 2 weeks. Coffee farms in Nicaragua, Honduras, and Guatemala were destroyed, decreasing the yields for the season. The hurricanes also caused severe infrastructure damage, and roads, bridges, and farm infrastructure, disrupting the harvest, processing, and transportation of coffee.

In Central America, coffee provides employment to the rural population. Climate change impacts are also a factor contributing to the outmigration of people from the region [19, 20]. Because coffee employs workers permanently but also temporarily, these workers are more prone to find employment alternatives elsewhere and some people abandon the region altogether. Permanent labors are more likely than temporary farmworkers to be formally registered, earn minimum wage, and have some degree of social security coverage, which is required by law SCA [21].

4. The coffee value chain in East Africa

In East Africa, small-scale farmers play a great role in coffee production. In Ethiopia, up to 70% of the total production is in the hands of smallholder farmers [22] and in countries like Rwanda and Burundi, most producers are small scale. As mentioned previously, farm gate prices are a major concern for farmers, as market prices locally and internationally are low and rarely cover the production costs. Even for farmers in cooperatives this remains a major challenge in the sector. In addition, there are infrastructural challenges affecting farmers. For farmers in remote areas, it is a challenge to bring coffee to collection centers, especially in areas with poor roads and infrastructure. For land-locked countries like Rwanda, Uganda, and Burundi, exporting coffee becomes more challenging, as lories must move the coffee through other countries, either Kenya or Tanzania, before it reaches the port for shipment. Again, the lack of proper roads and infrastructure is a major hurdle.

Market prices and fluctuations in international prices are felt by small-scale farmers in countries that depend highly on coffee exports as a source of income, such as Burundi, Uganda, and Honduras. Low farm gate prices coupled with higher production costs have resulted in losses or unsustainably low earnings [23–25]. Smallholder farmers lack the experience and knowledge to mitigate the price risks and are unable to cope with market fluctuations.

In East Africa as in Central America, coffee farmers are in the most vulnerable position in the value chain. Low farm gate prices and price volatility affects them severely. The C-price, along with certain factors that lead to great variability, influence farm gate prices in producing countries, and farmers cannot exert any influence or change these prices. Agricultural policies implemented by governments (i.e., coffee funds, subsidies) can make a huge difference in producers' livelihoods [26], but evidence suggests that paying a living income can improve the situation in the long term [27].

Development agents, NGOs, and other entities are actively promoting the transformation in the sector by advocating for the payment of a living income to farmers, particularly coffee farmers in East Africa. Living Income is the net annual income required for a household in a particular place to afford a decent standard of living for all members of that household. Just like the living wage in the garment sector, the living income discussion is gaining momentum globally among donors, NGOs, and governments. Therefore, living income is a critical issue that needs to be addressed in depth and discussed to find solution. Earning a living income means that the farmer will be able to provide basic but decent food to the household, provide basic decent housing according to the area standards, and provide basic health care and basic education for the children.

East African coffee farmers and their families live well below recognized living income benchmarks. Prices are a major issue affecting farmers, as the long-term nominal price of coffee in the futures market has remained roughly the same for over 40 years, resulting in a significant decrease in farmers' purchasing power. High population densities in smaller countries like Rwanda and Burundi mean that there is a continuously decreasing farm and plot size. Fragmentation of farm plots result in lower incomes for farmers [28].

5. Promoting sustainability and reducing inequalities in the coffee value chain

Sustainability in the coffee sector should not only be approached from an environmental perspective, but also an economic and social perspective. The farmers' livelihoods need a stable source of income for proper investments to be made in good agricultural practices which will ensure higher productivity and sustainability in the long run. This is only possible with a fair and stable income which allows the farmers to have the necessary resources to make investments. The discussion around living incomes for farmers in commodities like coffee have the objective to promote a decent standard of living. A decent income allows for social targets such as reducing child labor to decrease. In addition to these economic and social benefits, living incomes allow for environmentally detrimental practices such as deforestation to be reduced, because an economic viability is the catalyst for the sustainability of the entire coffee sector (Coffee Development Report, 2019).

Climate change hotspots and changing climatic conditions will render some coffee-growing regions no longer suitable [29, 30]. Global climate models used to explore three climate scenarios show that highly and moderately suitable Arabica coffee-growing environments around the world will decrease by 50% and 30%, respectively, by 2050 [30]. Central America is one of these areas where coffee production will be affected [31]. Without investment and a secure flow of economic resources in the sector, climate change mitigation strategies are impossible to reach for most small-scale coffee farmers in Central America and East Africa.

Coffee farmers will need to adapt to less predictable growing conditions and more extreme climatic events. Climate resilience requires the right investments at the right time, as coffee cultivation requires long-term planning. Most of today's production comes from smallholder farmers with fewer resources to cope with shocks, making them more vulnerable to climate change. Some countries have implemented strategies to mitigate climate change effects in their coffee sectors. Important adaptation measures implemented include planting improved coffee varieties or hybrids that have a higher drought tolerance or adopting more sustainable farming practices, like shade-grown coffee to reduce temperatures or mulching to maintain soil moisture [32]. The adoption of voluntary standards and certifications is another sensible strategy for farmers. Most farmers producing certified coffee must adopt good agricultural practices, and in most cases, they are trained by cooperatives, private sector actors, and other entities to comply with the specified practices to maintain the label. Improved practices often result in higher yields, which is beneficial to farmers. Certified farmers are also paid a premium for their coffee, which eases some of the economic pressures most of them face when making farm-level investments [27].

A critical factor to keep in consideration when addressing the farm-level sustainability of coffee farmers is the fact that 75% of the global coffee production is exported to international markets. The total generated value was US$20 billion on average in the period 2015–2020 [3]. Because most coffee is exported as green unprocessed coffee beans, farmers capture less value, while roasters and retailers in importing countries capture the largest share of the value addition. The Coffee Baraometer (2020) estimates that the average green coffee export value accounts for less than 10% of the US$200 to US$250 billion of revenues generated in the coffee retail market. In 2020, coffee prices dropped to 30% below the average price level reported over the past 10 years [11]. A study conducted in 13 countries indicates that coffee producers' average annual income decreased significantly in 2017–2018 [9]. The proportion of farmers living below the extreme poverty line of US$1.90 per day has increased significantly, and in countries like Nicaragua, by as much as 50%. Such low prices impact farmers, and coffee production is not economically viable for most coffee farmers. For the "typical" small-scale coffee production system in East Arica and Central America with lower possibilities of investment, the impact of low prices is high. For high-productivity countries where the more productive farms remain profitable, these fluctuations in prices have a negative impact, but the situation will never be as grave as it is for the smaller-scale, resource-poor farmers who cannot make the very basic investments. An example of the effect can be explained as follows:

"In 2019, with an average price of US$1.80 per kg of green coffee, a Colombian producer with 4.3 hectares coffee land could only reach a living income with a production of 1.46 metric tons per hectare [33]. Even under a very optimistic scenario with a simultaneous increase in yield and farmgate prices, a producer with a small plot of land would not reach an income level above the poverty line [34]. The development of economically viable coffee production is vital to many countries' efforts to combat extreme poverty [35]. Although Africa accounts for about 10 percent of global coffee production, these figures understate the importance of coffee production in terms of its contribution to a country' GDP, rural employment, tax incomes and export earnings. For instance, in Ethiopia and Burundi coffee is the largest single exported product by percentage in 2019" [3].

For coffee production to remain viable in the long term, these inequalities must be addressed. In both Central America and East Africa, smallholder coffee farmers do not have the means to adopt critical measures to mitigate climate change impacts or cope with an international market system which offers them little to no opportunity to have a profitable coffee production system. Sustainability can only be reached by minimizing the inequalities in the sector, and this can only be achieved by channeling more monetary resources to farmers through a transparent and fair pricing system.

6. Conclusion

Coffee is an important agricultural export for countries in Central America and East Africa. It is a sector which provides critical export revenues for the economies of these countries and employment in the rural sector. The numerous challenges smallholder coffee farmers face must be addressed in a coherent way to guarantee the sustainability of coffee as a global value chain. Because the governance structure of this value chain gives producers in developing countries no opportunity to influence or improve a situation such as market price, there only possibility to reduce inequalities and mitigate further risks associated with low incomes is through a combination of public and private sector initiatives to ensure a better distribution of gains from farm to retail. Market-driven initiatives such as sustainable sourcing and the implementation of voluntary sustainability standards can have a better effect if the right public policies and regulatory options are complimentary. Otherwise, a redistributive effect of profits trickling down to farmers is unlikely.

It is imperative to reduce the inequalities in this business for coffee to be sustainable. Better prices paid to farmers can help them cope with climate change impacts which will likely become a more critical issue in Central America and East Africa. Institutional coordination at the country level is important to help coffee farmers adapt and cope with the challenges mentioned in this chapter. Key stakeholders such as research organizations, local government institutions, producer associations, traders, exporters, buyers, and civil society organizations all have a specific role and sphere of action to address the needs in the sector. Through coordinated efforts among different stakeholders in the value chain, necessary changes can be implemented. The adoption of climate change mitigation strategies will only be successful through engagement among all stakeholders because smallholder coffee farmers cannot achieve these goals on their own.

Author details

Ingrid Fromm
School of Agricultural, Forest and Food Sciences, Bern University of Applied Sciences, Switzerland

*Address all correspondence to: ingrid.fromm@bfh.ch

References

[1] Expert Market Research (EMR). Global Coffee Market Outlook. 2020. Available from: https://www.expertmarketresearch.com/reports/coffee-market

[2] Food and Agriculture Organization of the United Nations (FAO). The State of Agricultural Commodity Markets 2020. Agricultural Markets and Sustainable Development: Global Value Chains, Smallholder Farmers and Digital Innovations. Rome: FAO; 2020

[3] Panhuysen S, Pierrot J. Coffee Barometer. Brussels; 2020

[4] Fairtrade Interantaional. Coffee. 2022. Available from: https://www.fairtrade.net/product/coffee

[5] Food and Agriculture Organization of the United Nations (FAO). Commodities in Focus: Coffee. 2022. Available from: https://www.fao.org/markets-and-trade/commodities/coffee/en/

[6] International Coffee Organization (ICO). Coffee Market Report February 2021. ICO Statistics. 2021. Acailable from: https://www.ico.org/documents/cy2020-21/cmr-0221-e.pdf

[7] International Trade Centre (ITC). The Coffee Guide. Geneva: ITC; 2021

[8] Smith J. Coffee landscapes: Specialty coffee, terroir, and traceability in Costa Rica. Culture, Agriculture, Food and the Environment. 2018;**40**(1):36-44

[9] International Coffee Organization (ICO). Identifying Coffee Sector Challenges in Selected Central American Countries and Mexico. Projects Committee Brief. London; 2018. Available from: http://www.ico.org/documents/cy2017-18/pj-120e-challenges-central-america-mexico.pdf

[10] Meister E. New York City Coffee: A Caffeinated History. Cheltenham: The History Press; 2017

[11] International Coffee Organization (ICO). Daily Coffee Prices. ICO Statistics. London; 2020. Available from: http://www.ico.org/coffee_prices.asp

[12] World Bank Group. Commodity Markets. Market Prices. Washington, D.C.; 2023. Available from: https://www.worldbank.org/en/research/commodity-markets

[13] Bacon CM, Méndez VE, Gliessman SR, Goodman D, Fox JA, editors. Confronting the Coffee Crisis: Fair Trade, Sustainable Livelihoods and Ecosystems in Mexico and Central America. Cambridge, MA: MIT Press; 2008

[14] Caswell M, Méndez VE, Bacon CM. Food Security and Smallholder Coffee Production: Current Issues and Future D. Burlington, VT: Agroecology and Rural Livelihoods Group, University of Vermont; 2012

[15] Lewin B, Giovannucci D, Varangis P. Coffee Markets: New Paradigms in Global Supply and Demand. World Bank Agriculture and Rural Development Discussion Paper No. 3, 2004. p. 150

[16] Watts C. A Brewing Storm: The Climate Change Risks to Coffee. The Climate Institute; Sydney; 2016. p. 16

[17] Rigaud KK, de Sherbinin A, Jones B, Bergmann J, Clement V, Ober K, et al. Groundswell: Preparing for Internal Climate Migration. Washington, DC: The World Bank; 2018

[18] IHCAFE. Los retos de la caficultura hondureña ante el COVID-19. IHCAFE Webinar. 2020

[19] Angel M, Palencia G, Menchu S. Coffee Crisis in Central America Fuels Exodus North. Reterus; 2021. Available from: https://www.reuters.com/markets/commodities/coffee-crisis-central-america-fuels-record-exodus-north-2021-12-08/

[20] Reichmann D. Putting climate-induced migration in context: The case of Honduran migration to the USA. Regional Environmental Change. 2022;**22**:91

[21] Specialty Coffee Association (SCA). Farmworkers and Coffee: The Case for Inclusion. An SCA Whitepaper. 2018. p. 23

[22] Abate G, Bernhard T, Regass B, Minten B. Improving Coffee Productivity in Ethiopia: The Impact of a Coffee Tree Rejuvenation Training Program on Stumping. 2021

[23] Estrella A, Boucher S, Sanger C. Profitability of Coffee Farming in Selected Latin American Countries – Interim Report. London: International Coffee Organization; 2019

[24] Food and Agriculture Organization of the United Nations (FAO). Food Outlook: Biannual Report on Global Food Markets. 2018

[25] Meier C, Sampson G, Larrea C, Schlatter B, Voora V, Dang D, et al. The State of Sustainable Markets 2020: Statistics and Emerging Trends. ITC, Geneva; 2020

[26] Grabs J. The lowest coffee price in a decade. Food (Policy) For Thought. 2018. Available from: https://foodpolicyforthought.com/2018/08/23/the-lowest-coffee-price-in-a-decade/

[27] Bermudez S, Vivek Voora V, Larrea C. Coffee prices and sustainability. Sustainable Commodities Marketplace Series. 2022

[28] Cordes KY, Sagan M, Kennedy S. Responsible Coffee Sourcing: Towards a Living Income for Producers. 2021. Available from: https://scholarship.law.columbia.edu/sustainable_investment_staffpubs/199

[29] Furer M, Jurt C, Fromm I, Malek Z, Kuonen L, Marchesi S, et al. Fairtrade and Climate Change: Systematic Review, Hotspot Analysis and Survey. Bonn: Fairtrade International; 2021

[30] Grüter R, Trachsel T, Laube P, Jaisli I. Expected global suitability of coffee, cashew and avocado due to climate change. PLoS One. 2022;**17**(1). Article e0261976

[31] Fromm I. Building resilient value chains after the impact of the COVID-19 disruption: Challenges for the coffee sector in Central America. Frontiers in Sustainable Food Systems. 2022;**5**:775716

[32] Pham Y, Reardon-Smith K, Mushtaq S, Cockfield G. The impact of climate change and variability on coffee production: A systematic review. Climatic Change. 2019;**156**:609-630

[33] Solidaridad. Costos de producción de café, 2019 Colombia. 2020. Availble from: https://www.solidaridadsouthamerica.org/sites/solidaridad southamerica.org/files/publications/200607_informecostos_2019.pdf

[34] TFCLI. Strategy handbook, a fact-based exploration of the Living Income Gap to develop effective sourcing and pricing strategies that close the gap. 2020. Available from: https://www.

idhsustainabletrade.com/publication/strategy-handbook-full-report/

[35] CABI. Stakeholders launch a $950m initiative aimed at boosting productivity for the coffee industry. CABI News. 2019. Available from: https://www.cabi.org/news-article/stakeholders-launch-a-950m-initiative-aimed-at-boosting-productivity-for-the-coffee-industry/

Chapter 7

Value Chain Implementation in Rural-Scale Integrated Coconut Farming System in North Sulawesi Province, Indonesia

Agustinus N. Kairupan, Jantje G. Kindangen, Gabriel H. Joseph, Ronald T.P. Hutapea, Ibrahim Erik Malia, Paulus C. Paat, Derek Polakitan, August Polakitan, Jefny B. Markus Rawung, Meivie Lintang, Joula O.M. Sondakh, Payung Layuk, Olvie Grietjie Tandi, Herlina N. Salamba, Nelson H. Kario, Jonathan Anugrah Lase and Rindengan Barlina

Abstract

The value of coconut production in North Sulawesi reaches almost 10% of Indonesia's total coconut production, and more than 95% is managed by farmers. The main problem in coconut development is that the management of coconut cultivation is still traditional, and the quality of the products produced is still low, so that coconut commodities that are multipurpose relatively have no added value. The economic value of coconut is still very dependent on primary products, namely coconut seeds and copra. To increase farmers'income, the solution that can be taken is to implement the value chain in integrated coconut farming by diversifying coconut products. This is a review chapter that aims to present the implementation of value chain analysis approach and formulate the strategic development of integrated coconut farming system on a rural scale in the North Sulawesi province. Implementation of the value chain by diversifying coconut commodities can provide added value. By managing an integrated coconut farm that produces 3–5 processed products, the economic value of coconut will increase by about 5–6 times. Several issues become obstacles in the development of the value chain, including the lack of farmer participation in the modern chain, weak market power and market access, weak contractual relationship strength, and weak agricultural sector development orientation in the provision of employment.

Keywords: value chain, agriculture, coconut, diversification, income

1. Introduction

The role of the agricultural sector in Indonesia contributes significantly to development and the national economy. It absorbed 29.769% of the total labor force in Indonesia and contributed 9.85% to the national GDP in 2021 [1]. To realize advanced, independent, and modern agriculture as the ideals of Indonesia's agricultural development in the future to improve the economy and welfare of farmers, the priority aspect is to increase productivity and added value of agricultural products.

In the midst of increasing activity of foreign consumer good import flows, agricultural commodities, especially plantation sub-sector commodities, in Indonesia are still the leading commodities in the international world. Among them is the coconut plant, which is a plantation plant with the largest area in Indonesia. The total area of coconut in 2018 reached 3,417,951 hectares, of which about 99% or an area of 3,385,085 hectares was cultivated by smallholder plantations, cultivated in monoculture or overlapping with other crops; large state plantations covered an area of 3482 hectares or 0.11%, and large private plantations covered an area of 29,024 hectares or 0.85%, with national coconut production reaching 2.85 million tons in 2021 [2].

The North Sulawesi province is one of the coconut production centers in Indonesia with a coconut plant population in 2018 recorded over an area of 273,331 hectares and is the second largest area after Riau province [2]. The value of coconut production in North Sulawesi reaches almost 10% of the total Indonesian coconut production, and more than 95% is managed by farming communities; until now, the acquisition of value from various products has been limited to primary products. Therefore, farmers'income from the business sector is still relatively low, not comparable to the potential resources available to produce various competitive and economically valuable products.

The main problem is that generally the farming community finds it difficult to get out of the condition of product value acquisition; farmers rely only on the value of primary products. Farmers who practice coconut monoculture and produce copra or coconut granules as their main product can obtain a net profit of Rp 9,000,000 per hectare per year or around Rp 750,000 per hectare per month. The productivity potential is approximately 2 tons of copra per hectare per year or around 8,000 coconuts, with a copra selling price of around Rp 10,000 per kilogram. This condition has an impact on the existence of farmers who are becoming increasingly helpless, and there tend to be more and more coconut farmers who become poor. Allorerung et al. [3] reported the results of research in North Sulawesi and Riau, which showed that the socioeconomic level of coconut farmers is below the poverty line, which is less than US$ 200/capita/year.

To overcome the problems mentioned above, initially, the farming community must get out of the discomfort zone by empowering all existing potential through a process that will face various obstacles, so a persistent, resilient, and confident attitude is needed. The success of each farm is highly dependent on the ability of each farmer in the management of the farm. Hernando [4] stated that agricultural management is the ability of farmers to determine, organize, and coordinate the factors of production that are best controlled and able to provide agricultural production as expected. It is further stated that the measure of management success is the productivity of each factor and the productivity of its business.

In the next few years, demand for various coconut products is expected to increase. To anticipate this, increasing the income of coconut farming needs to be done through the value chain approach. Value chain analysis is used as a strategy analysis tool to better understand competitive advantage, where a manufacturer or firm can increase added value or lower costs to make the business more competitive [5]. This chapter is a review that aims to present the implementation of value chain analysis approach and formulate the strategic development of integrated coconut farming system in the North Sulawesi province.

2. Methodological approach

This chapter uses a research design with a scientific review method. This method is used to conduct descriptive exploration and analysis of data on the topic discussed, which are sourced from various scientific references, such as journals, books, proceedings, electronic and print media, and other relevant published sources. The steps of the data analysis process are uniting and equalizing into complementary data. Collecting various data obtained is described descriptively to find problems and solutions for future development. The results of the identification and description of the data are presented in the form of charts or flowcharts.

3. Results and discussion

3.1 Concept and stages of implementation of value chain analysis

The value chain approach can help to understand how trends reshape the value chain, identify who is leading and addressing them, answer broad and specific questions, and take a relationship/friendship approach. Poole [6] states that the value chain concept offers an analytical approach to explore market-agriculture-food relations, to assess potential cross-sectoral contributions, and to identify incentives and barriers in production and consumption.

Designing value chain analysis, among others, by improving the ability of local companies to compete, increasing returns for consumers, encouraging targeted policies, and empowering the participation of farmers in rural areas to engage in all productive activities. Stringer [7] stated that value chain analysis is one of the concepts of how to add activities and enlarge the value of products to the maximum in the supply chain. Value chain analysis offers the opportunity to maintain competitive advantage through a relational approach. Value chain analysis focuses on end users and chain organizations to meet the needs and desires of consumers [8].

The main objective of value chain management is to increase or maximize profits and minimize financing. Value chain analysis always refers to how much and how long it takes to prepare or present goods for sale. Creating jobs and local income always refer to the value chain with an understanding of what the market will pay [7].

The amount of value of a product depends a lot on the outside of the system, especially knowing what exactly the needs and desires of consumers are. Kotler and Keller [9] states that marketing and consumer behavior are derived from the marketing concept that states that the essence of marketing is meeting the needs of consumers, creating value that fits their needs, and retaining customers. To obtain the value of the final product in marketing, the dominant result is determined by the degree to which the producer knows the consumer's desire for a product and a variety of products.

The value chain provides a vehicle for identifying ways to create differentiation through value development [10]. Value chain activities are categorized into primary activities and supporting activities. Supporting activities bring together functions further dividing the specific main activities in the value chain would be beneficial.

To realize the maximum product value, it is necessary to know what the needs and desires of consumers are; then, the maximum economic value is more focused on the desires and needs of consumers. The development of smallholder oil palm plantations will generate maximum product value and income if sub-systems in the distribution channel chain can be created/added to various productive activities and if all of them are interrelated and strengthened.

Several problems become obstacles in the development of the value chain, including the lack of participation of farmers in the modern chain, weak market power and marketing access, the absence of certification, weak strength of relationships through contractual agreements, and weak orientation of regional development in the agricultural sector in the provision of employment. Some of the fundamental challenges for the main actors include not believing in payment, not being able to supply inputs because working capital is limited, and not being able to supply input quality, technology, and management. The solution should be to build economic and social institutions for farmers, such as through a combination of farmer groups oriented to developing business partnerships based on mutual agreements, negotiating to rebuild trust in order to produce sustainably with timely payment systems, escorting input allocation, and escorting management and technology.

The stages of value added analysis activities are as follows: (1) present a real problem situation; (2) analyze the condition of the situation with stakeholders by analyzing the inventory of value chains, institutions, and social systems; (3) organize the sequence of problems encountered; (4) make improvements and solutions to existing problems through system identification, model design, and improvement suggestions in several stages of testing; (5) distinguish between the model designed with the real situation in the field; (6) systematic changes in desires that allow, and (7) act/implement activities to change things to be more productive to obtain added value.

Adding value through process innovation can be done, among others, via a good service system, looking different, promotion, building distribution channels, doing some marketing methods, transportation management, storage systems, cooling systems, quality assurance and insurance, application of production technology, post-harvest handling, cleaning, harvesting techniques, application of culture, use of good plant materials, organization of the growing season, certification, and labor management. Adding value through product innovation can be done in the form of product attributes, originality information, appropriate ingredients, mixtures and packaging boxes, packaging forms, product identity explanations, health information, yield quality, use of appropriate plant varieties, appropriate timing, consistency of product provision, and product volume.

3.2 Application of value chain In oil palm cultivation

The coconut crop in the North Sulawesi province is one of the leading commodities of the region, which has the largest area and makes up the main source of income of the farming community and contributes dominantly to the economic growth of the region. The reality of the field shows that factually it is not directly proportional to the living conditions of the farming community as coconut owners.

The production of coconut farming systems is sold by most farmers only in the form of coconut or copra granules. By relying on coconut products, the economic value of coconut is mostly only enjoyed by intermediaries, especially producers. The incommensurate acquisition of added value gives an idea that the coconut

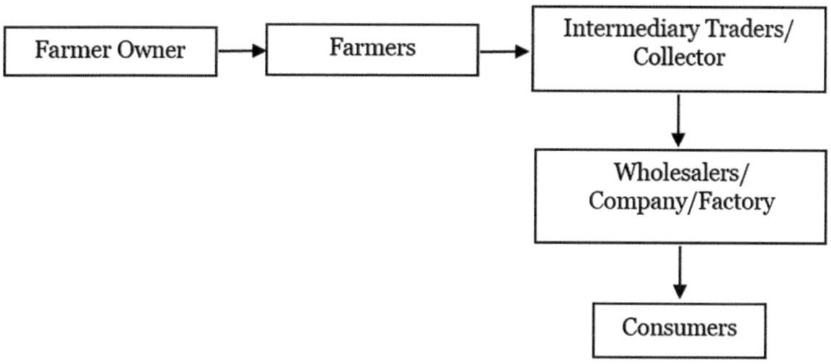

Figure 1.
Flow of activity of actors in the copra value chain in North Sulawesi.

agribusiness system in this area is still conditioned to its spread [11], which is characterized by weak or no functional relationship between each level of business. In this case, farmers only become suppliers of raw materials, and intermediaries/producers accommodate primary products for further processing or distribution to other parties for maximum profit.

In general, coconut farmers sell individual coconut products directly to traders. Until now, the coconut marketing chain produced by farmers has been quite long, so there needs to be an effort to take efficient coconut marketing actions without going through some unnecessary intermediaries. The activity flow of copra value chain actors from farmers in North Sulawesi is shown in **Figure 1**.

Price difference in the form of copra products between village traders and wholesalers or manufacturers is usually around Rp 300-Rp 400/kg, and in the form of granular products, there is a difference of around Rp 100-Rp 150/kg. If farmers have as much as 4000 kg of copra/year, then every year, there is an opportunity for additional income through a collective sales system. For sales like this, each farming family will have an additional income of around Rp 1.2-Rp 1.6 million/year for copra and Rp 1.6-Rp 2.4 million/year for sand coconut. Assuming in 1 village, there is a copra equivalent production of 800 tons, by implementing a collective sales system, there will be an increase in revenue of Rp 240-Rp 320 million/village/year for copra and Rp 320 million-Rp 480 million/village/year for coconut sand.

Field facts to date show that the mainstay of obtaining the value of coconut products is still very dependent on primary products in the form of coconut granules or copra [12]. The shape and color of copra products sold by coconut farmers are still the same as the shape and color of copra about 30–40 years ago. To overcome the low income of farmers, it is necessary to diversify and create the value of processed coconut products to increase the value of processed coconut products so that farmers not only focus on selling coconut as a primary product but also can process it into other products that have high economic value and will ultimately have an impact on increasing farmers'income. Diversification is the diversification of products to meet consumer needs.

Some forms of coconut processing products that can provide added value include (a) white copra, (b) virgin coconut oil (VCO), (c) nata de coco, (d) shell charcoal, (E) vinegar acid, (f) coir fiber, and (f) application of diversification efforts in coconut plantation areas with other types of crops or livestock. Diversification and value creation of processed coconut products on a rural scale is seen in **Figure 2**.

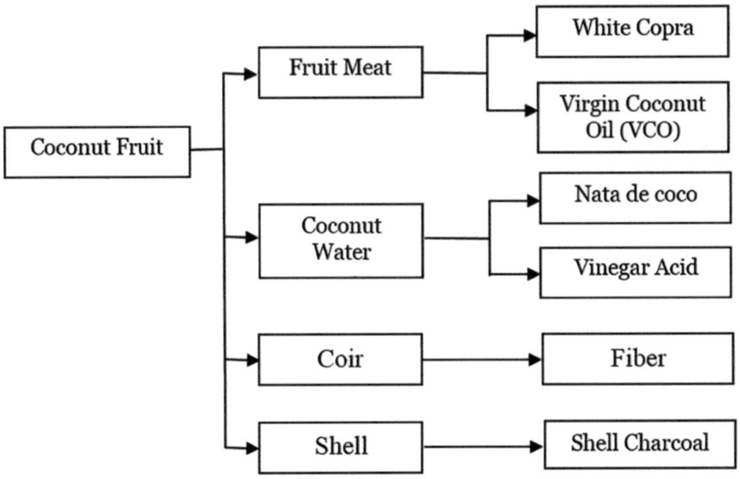

Figure 2.
Diversification and value creation of processed coconut products on a rural scale.

3.2.1 White copra

White copra is a type of quality copra produced from fresh fruit pulp through a drying/organizing process at a temperature of 60–75°C; and has low moisture content of between 5 and 6%, free from boletus (mushrooms); and has a much whiter and cleaner color. By applying coconut processing into white copra products, assuming that every farmer has a coconut with an average product of 4 tons/year equivalent to copra, there can be an added value of the price difference with ordinary copra of Rp 700–Rp 1000/kg. By implementing a white copra processing system, each farmer has an additional income opportunity of around Rp 2,400,000-Rp 4000,000/year. If a village produces 800 tons of copra per year, they can increase their income by approximately Rp 560,000,000 to Rp 800,000,000 per year by processing the copra into white copra.

Wardanu and Uliyanti [13] reported that the result of his research to obtain the value of efficiency or Revenue Cost Ratio (R/C) was 1.61, which means that every Rp 1 cost incurred for the processing of white copra will generate a gross income of Rp 1.61 and a net profit of Rp 0.61. Ha shows that the business of processing white copra is efficient or commensurate with the company, where the income obtained amounted to Rp 5,650,000.00/production. At the same time, the total costs incurred amounted to Rp 15,929,433. 33 per production to obtain a profit of Rp 9,720,566. 67/ production.

3.2.2 Virgin coconut oil (VCO)

Virgin coconut oil or better known as (VCO) is oil produced from the purification of coconut milk by a certain method to separate the chemical elements gradually [14]. If 25% of coconut products in a village can be processed into pure coconut oil (VCO), with an equivalent of 200 tons of coconut production per year, approximately 6,500 liters of VCO can be produced, worth around Rp 1.85 billion per year per village at a price of Rp 30,000 per liter. The potential difference in value is quite large compared to just making copra worth more than Rp 1 billion/year/village.

The results of the financial feasibility analysis showed that the BCR value of the VCO processing industry was greater than 1 (2.08–2.28), as well as the Net Present Value (NPV) was positive. This result illustrates that the amount of net income is much greater than the amount of expenses, or it is profitable to strive for. Meanwhile, the Internal Rate of Return (IRR) value shows a number higher than the discount factor set, namely 12%, 15%, and 20%. Under the conditions of prices and products specified in the analysis, IRR values of 35.5% and 39.4% or greater than 20% are obtained, which means that the investments allocated to the business of this industry will provide the maximum net profit, assuming guaranteed marketing continuity. The value of coconut production through the implementation of this business reached around Rp 1665/grain, and farmers'net profit reached Rp 955/grain.

3.2.3 Nata de coco

Nata de coco is a food product made from coconut water. Coconut water is waste from coconuts, potentially generating added value after intervention with treatment technology. The results of the financial feasibility analysis showed that the Benefit Cost Ratio (BCR) value of Nata de coco processing industry was greater than 1 (2.32–2.48), as well as the NPV value was positive. This result illustrates that the amount of net receipts is much higher compared to the amount of expenses, or it is profitable to strive for. Meanwhile, the IRR value shows a number higher than the discount factor set, namely 12%, 15%, and 20%. In the conditions of prices and products specified in the analysis, obtained IRR of 35.3% and 39.7% or greater than 20% means that the investment allocated to the industrial business provides the maximum profit, assuming the marketing of the resulting product is guaranteed. The implementation of this business obtains additional value of coconut production of around Rp 575/grain, and farmers obtain additional net profit of RP 350/grain.

Some results of the study reported that the calculation of value-added analysis of processing Nata de coco with the use of coconut water as much as 400 liters of Rp 350/liter produces Nata de coco of 167.73 kg, with a selling price of Rp 15,000/Kg [15]. Furthermore, it was stated that the average income in the Nata de coco processing business by the household industry was economically profitable with an income level of Rp 2,159,445.

3.2.4 Shell charcoal

Shell charcoal is charcoal made by carbonization of coconut shell/shell [16]. In North Sulawesi, a copra-producing region, coconut shells are found in large quantities, which is the result of copra making. During this time, the shell was used as fuel in the manufacture of copra, but most of the rest was wasted. The huge potential of shell production needs to be utilized for productive activities to increase its added value because coconut shell processing technology is relatively simple and can be implemented by small businesses [17].

The economic value of coconut shell can be increased by using it as an industrial raw material, for example, as a raw material for making activated charcoal. Activated charcoal is indispensable in the cooking oil processing, sugar, and other food and non-food industries, because of the ability of activated charcoal to absorb color, gas, and other impurities [18]. With the increasing need for activated charcoal, the need for coconut shell charcoal, which is the raw material for making activated charcoal, will also increase.

The results of economic analysis showed that the BCR value of shell charcoal processing industry is greater than 1 (1.28–1.33), as well as a positive NPV value. This result illustrates that the amount of net receipts is higher than the amount of expenses, or it is profitable to strive for. Meanwhile, the IRR value shows a number higher than the discount factor set at 12%, 15%, and 20%. In the conditions of prices and products determined in the analysis, the obtained IRR value of 36.2% and 39.7% or greater than 20% means that the investment allocated to the business of this industry will provide substantial benefits, assuming the marketing of its products is guaranteed. The application of this business generates additional value of coconut production of around Rp 85/grain, and additional net profit of Rp 22/grain is obtained by farmers.

The results of several studies show that the average total cost incurred in the coconut shell charcoal business is Rp 2,8.34.586,57/production; the average income received by coconut shell charcoal businesses is Rp 5,748,750,00/production. Meanwhile, the average income is Rp. 2,914,163.43/production. The value of RCR (Return Cost Ratio) in the manufacture of coconut shell charcoal is 2.02. This indicates that the effort is feasible (RCR > 1). The added value for charcoal shell making business is Rp. 799.02/Kg [19].

3.2.5 Vinegar acid

Food processing derived from coconut water is processed not only into Nata De coco but also in the form of vinegar acid. One country that uses the raw material of coconut water in the manufacture of vinegar is the Philippines by applying a fermentation process to it.

Vinegar or acetic acid is a liquid that tastes sour and is made through the process of alcoholic fermentation. Acetic acid is commonly used for household purposes as a flavoring agent in food. Acetic acid can be prepared from ethanol-containing substrates, which can be obtained from a wide variety of materials, one of which is coconut water. In North Sulawesi, the availability of coconut water in abundance is a potential that has not been fully utilized. Coconut water can be made into vinegar traditionally by the community through fermentation techniques.

The analysis showed that the industrial BCR value of vinegar acid was greater than 1 (2.1–2.8), as well as a positive NPV value. This result illustrates that the amount of net income is much greater than the amount of expenses, or it is profitable to strive for. Meanwhile, the IRR value shows a number higher than the discount factor set, namely 12%, 15%, and 20%. In the condition of product prices determined in the analysis, IRR values of 36.5% and 39.5% or greater than 20% are obtained, which means that the investment allocated to this industrial business will provide maximum net profit, assuming the continuity of marketing of its products is guaranteed. The application of this business can obtain additional value of coconut production of around Rp 760/grain and additional net profit of Rp 500/grain for farmers.

3.2.6 Coir fiber

Coconut fiber is a by-product of coconut processing products derived from the separation process of fruit-skin fibers. The content of coir fiber in coconut is a fairly large part, which is 35% of the overall weight of coconut. Each coconut grain contains an average of 525 grams of fiber (75% coir) and 175 grams of cork (25% coir).

The results of observations in the field show that coconut fiber is mostly discarded at the time of manufacture of copra, or left alone, to rot and dry under coconut stands. Manikantan et al. [20] argued that the coconut belt still has a fairly good economic value. Coconut fiber can be obtained. Coir fiber for coconut-producing neighboring countries has become an export commodity that supplies the world's needs. With this phenomenon, it is increasingly clear that this fact is a great potential for the development of the coir management industry, which is then expected to generate increased income, which has a positive effect on increasing the value of companies that can support the lives of lower- and middle-economic communities that use coconut coir as their livelihood [21].

The results of economic analysis showed that the BCR value of the coir processing industry >1 (1.60–1.69), as well as the NPV value was positive. This result illustrates that the amount of net income is much greater than the amount of expenses, or it is profitable to strive for. Meanwhile, the IRR value shows a number higher than the discount factor set at 12%, 15%, and 20%. At the price and product conditions determined in the analysis, the obtained IRR values of 34.6% and 39.8% or greater than 20% mean that the investment allocated to the business of this industry will provide substantial benefits, assuming the continuity of product marketing is guaranteed. The application of this business to obtain additional coconut production generates value of about Rp 285/grain and additional net profit of RP 122/grain for farmers.

Applied to an integrated coconut industry business like this, when farmers have an area of 1 ha of coconut with copra equivalent productivity of 2 tons/ha/year or 8000 grains/ha/year, farmers get a net profit of Rp 12,000,000, compared with the present condition of grain sales, where they only get a net profit of Rp 1.800.000/ha/year. By building an integrated coconut industry business like this, the value of coconut production increases by about 5–7 times the value of primary production, industrial capacity of 1500 grains/day or 450,000 grains per year, or equivalent to 112.5 tons of copra/year.

The results of research [22] showed that the processing of coconut fiber into coconut fiber in the integrated agro-industrial business area (strong) in the District of West Coast could provide added value for the processor of Rp 1.890 for each kilogram of raw materials and provide an increase in added value of 57.55 percent.

3.2.7 Application of diversification efforts in coconut plantation areas with other types of crops or livestock

Diversification is the diversification of products to meet consumer needs. This effort is made to generate added value and increase sales [23]. With this, integration patterns, land production, and productivity can be increased, or farmers'income can also increase and be more resistant to various risks, such as seasonality, price, and income. Farmers obtained a net profit of Rp 9 million/ha/year or Rp 750 thousand/ha/month from monoculture coconut cultivation with copra primary products or granulated coconut with a productivity potential of 2 tons of copra/ha/year (approximately 8 thousand grains) at a copra price of Rp 10 thousand/kg. Coconut prices now fluctuate so much that when the price of copra is only Rp 5000 thousand/kg net, profit obtained is 4.5 million/ha/year or Rp 375,000/ha/month. This is in line with what Mahmud [24] stated that the coconut monoculture business only earns revenue of around Rp 2.5 million/ha/year or Rp 200,000/month.

Business development integration of coconut plants with livestock in essence optimally utilizes coconut crop area so that the feed available in quantity and quality is in accordance with the needs of the cultivated livestock. The more livestock that want to be cultivated, the more feed products produced per unit area of coconut farming. The results of observations in various areas of coconut centers where farmers cultivate cattle in coconut plantation areas, with traditionally only 2 cows, show that every year the value of income from livestock business exceeds the income from the value of coconut production (primary product).

Kindangen et al. [25] reported the results of their research on the feasibility of some business patterns of integration of coconut crops with livestock. In the integration business pattern, a coconut + corn + cow (3 parents) in the 1st year earned a profit of Rp 525,000, in the 3rd year Rp 11.285,000, in the 7th year Rp 15,005,000/ha/year, and in the 10th year Rp 24,450,000/ha/year. Business integration pattern b, coconut + corn + cow (8 parents) + grass king, in the 1st year obtained a negative profit above Rp 4000,000; in the 3rd and 5th year, it obtained a profit above Rp 10,000,000 and Rp 11,000,000/year, respectively. Furthermore, in the 7th and 9th year, the profit became more than 2 times from each has reached more than Rp 23,000,000/year. Most of the labor can be family labor, especially for the maintenance of livestock, so that family income becomes greater. In the 3rd to 5th year, the total family income reached above Rp 16,000,000/year, and in the 7th to 9th year, it reached above Rp 28,000,000/year. The business implementation is designed for a 10-year period, and during the 3rd to 5th year, the plan resulted in an added value that was almost four times greater than that of a coconut monoculture alone. By the 7th to 9th year, the added value had increased to more than seven times its original value. Overall, through integration with cattle, the added value is almost 2 times, starting from the 3rd to the 5th year, and starting from the 7th year, the added value reaches 3 times more than just the integration of coconut + corn.

The application of this business pattern contributes to the income from cattle business to be greater, with the proportion reaching 75%. In the integration business (pattern C) coconut + grass king + gamal/lamtoro, the early years of the period required considerable initial funds of more than Rp 80,000,000, especially due to investment for 20 cows/year, and starting from the 3rd year, a profit of more than Rp 28,000,000/year was reached, and this number was reached until the 5th year. Furthermore, starting from the 6th year, profits of more than Rp 50,000,000/year were obtained.

3.3 Value chain implementation prospects

The potential to obtain a minimum income value of about 3–5 times the existing conditions is very likely to be achieved through the development of processing activities of some products, and the exploitation of other crops and livestock in the coconut plantation area. The development of various productive business activities will be faster if supported by building and empowering economic and social institutions toward village-owned enterprises. The application of productive coconut farming that can produce diverse products coupled with the processing of various coconut products will gain added value for farmers. If the potential of coconut copra is equivalent to 800 tons/village as mentioned earlier, it means that during this time, the coconut farming community will have lost about ten billion rupiah/village. If only all the leading commodities in this area of the development system approach the value chain, then during this period in every village, coconut centers will have lost tens of billions of dollars. It must be realized that the result of this kind of helplessness is the main cause that more and more people who continue to pursue business in the agricultural sector become poor, even very poor.

The development of coconut farming through the value chain approach is identical to the business development of farming communities in rural areas in the future.

Therefore, if this business starts to be pursued now and is carried out consistently by all relevant parties and is sustainable and fully supported by all relevant institutions, then only, in a relatively short time, there will be a real improvement in the economy of the community and the region. With the establishment of economic institutions, including the seriousness of the government to facilitate it, it can be predicted that the condition of the farming community in the next 5.10–20 years is likely to be better. Dornan and Maxwell [26] argue that true success is determined by our attitude. The success we achieve depends more on the way we think than any other factor.

Success in the development of various agricultural commodities, the human factor being the most decisive in this regard, concerns the attitude of the main actors, especially the farming community itself and related supporting elements. In the early stages, the attitude of farmers' institutional managers is decisive for the achievement of success. The main thing that must be instilled in the institutional managers of agricultural communities is to respond positively to this institution as the main foundation for realizing the welfare of agricultural communities in the future. Formation of professional business people is a long process; its formation can only be done through practicing the business world through the processes of structuring, communication, and management systems and rational leadership, as well as a steady career plan [27].

Therefore, it is necessary for all members of an institution to learn and apply a positive thinking attitude when facing various challenges in the development of agricultural businesses with measurable results. As institutional managers, their responsibility at the initial stage is to create a simple business plan in their environment that will then be conditioned per the role of each actor to behave and behave positively so that their development is better in the future. The future of farmers in rural areas is expected to be about not only the economic growth of each farmer and region but also more than that. The future of farmers is expected to improve measurable welfare, realizing the guarantee of a decent farming family life in a sustainable manner.

The North Sulawesi province has several villages around 1500, and about 50–60% of them have coconut planting area, and future development should be such that each village can design the development of coconut farming through the value chain approach, at the initial stage, with the acquisition of added value of an average of Rp 300 million/village/year and then the acquisition of added value of all existing villages of Rp 450 billion/year or about 1.8–2% of Gross Domestic Product (GDP) at the prevailing price. The acquisition of added value from this amount will allow each village to gradually build a village-owned enterprise. In the 2004 regional autonomy law, it is stated that "villages can establish village-owned enterprises based on the needs and potential of the village" [28]. Therefore, any combination of existing farmer groups can design their income per farmer household in the short term (1–2 years), medium term (5 years), and long term (>10 years).

Through the presentation of value chain analysis, it is expected that the insight of the farming community will be open so that though farmer groups and farmer group associations immediately design it for each member in the next 5 years, a minimum income of Rp 50 million/ha/year will be obtained, and there will be a real increase in each period over the next 5 years. This will soon happen en masse if this effort becomes a program in every coconut center village that will be designed for development every 5 years along with the target type and number of products to be produced. The existence of the authority of regional autonomy should be the economic condition of each village where all the assets of the farming community need to be designed to achieve the target results for each year.

3.4 Strategic increase in the added value of coconut farming

Coconut farming, which is the mainstay of increasing farmers'income, in the future is faced with various challenges that hinder its development, so anticipatory steps are needed to develop effective strategies, namely operational steps that can maintain and even increase the added value of coconut farming.

Challenges that are predicted to arise in the development of coconut farming are related to the provision of raw materials for coconuts. Many coconut plants are now unproductive due to the relatively old age of the plant and disruption of plant pest organisms. This condition aggravates the achievement of coconut production, whose productivity is also not optimal because the coconut seeds planted are not superior varieties, but random seeds.

Many coconut farming systems still focus on primary products because the downstream industries that process coconut development are still limited. Coconut is known as a versatile tree that has not been managed optimally. The economic utilization of coconut is mainly focused on its fruit, shell, fiber, and sticks. This traditional approach reinforces conventional coconut farming activities. However, at the macro level, there are challenges due to the lack of government support for coconut farmers. Agricultural development priorities are still oriented towards staple food production, particularly rice.

Nasution and Rachmat [29] conducted research on aspects that hinder the development of oil palm related to market demand, lack of technological adaptation, and lack of capital controls. On the prediction of development challenges and constraints, coconut farming solutions include coconut cultivation, plant rejuvenation, and diversification of coconut processing. Based on these lessons, we developed strategies to improve added value of coconut products, which are differentiated according to upstream sub-sectors, aquaculture, and downstream sub-sectors.

In the upstream sub-sector, efforts to increase added value are focused on the introduction of superior seed use, provision of rejuvenation, rehabilitation, expansion, empowerment of farmer groups, provision of supporting facilities, improving the quality of post-harvest tools, and capital facilitation and initiatives to form farmer corporations. Local governments are expected to make efforts to trigger copra prices by developing the coconut cooking oil industry on the scale of farmer groups, namely by assisting farmer groups with coconut oil production machines and also through government efforts to improve the welfare of farmers [30, 31].

Anticipatory steps to maintain the continuity of coconut business development include replacing old plants and damaged plants and unproductive plants by rejuvenating plants. Although results are not obtained in the short term, at least for the long term, farmers are guaranteed to get a source of income from this coconut plant. To do rejuvenation takes approximately 4 years to produce again. Relatively long term concerns the needs and continuity of income. Therefore, while rejuvenating, farmers are encouraged to take advantage of vacant land with annuals as intercropping plants.

In a situation of decreasing prices of coconut products, including copra, efforts to reduce the risk of agricultural income through intercropping are an effective option, especially in monoculture coconut cultivation. Intercropping among coconut plants is supported by the results of Agronomic Research on monoculture plants whose land utilization is only 20 percent.

Other strategies to increase the added value of coconut farming in the upstream sub-sector are capital credit facilitation for intensification, rehabilitation, and rejuvenation; technical and institutional development of production; the existence of institutions such as "Coconut Board;" provision of Information Technology and market; improvement of the legal status of business land ownership; and infrastructure development.

Rejuvenation/rehabilitation of monoculture coconut plantations in a range of old/damaged/unproductive plant conditions, with superior varieties, through an integrative people's agricultural diversification system is based on coconut plants, with spacing arrangements (up to 16 M x 6 m) integrated with livestock. In the subsector of cultivation, efforts are made to intensify the control of plant pest organisms and farm diversification.

Diversification of coconut farming aims to anticipate the decline in coconut farming performance. When the condition of coconut marketing goes down, farmers are still saved economically because there is a harvest of intercropping plants cultivated in the alleys of coconut plants. Vacant land between coconut trees used for crops or short-lived annuals can provide income for farming households.

The practice of diversifying coconut businesses with food crops, industrial crops, and horticultural crops has been practiced by several coconut farmers in North Sulawesi. Besides being able to obtain added value from the sale of coconut plants, the existence of intercropping plants has proven effective in improving coconut production performance. In the downstream sub-sector, increasing the added value of coconut is done by diversifying the processing results, increasing the capacity of coconut farmers, and improving the performance of market promotion.

In addition, to ensure the sustainability of coconut cultivation, it needs the support of existing institutions such as farmer groups and community economic institutions (cooperatives). Agustinus et al. [32] stated that to improve the value chain, namely to strengthen farmer groups' existence to increase bargaining power and reduce transaction costs in marketing. Meanwhile, increased vertical coordination is carried out by building a network of partnerships with market participants and fulfilling contractual agreements in profitable markets. One of the institutional functions is as a vehicle for learning, cooperation, discussion, and capital.

The success of coconut cultivation is not only determined by technical factors. Aspects of knowledge, attitudes, and skills of the main actors and business actors are also determining factors. Therefore, increasing the capacity of the main actors and actors of coconut cultivation is very important. Coconut plantation management by farmers is strongly influenced by coconut market conditions. In market conditions that provide attractive prices, farmers generally manage coconut plantations well. The outpouring of family labor is aimed at the management of coconuts in the garden or field, since there is hope of obtaining an adequate income. However, when the market price of coconut plummeted, farmers do not pay attention to the coconut harvest. Another strategy is to encourage synergy in the development of coconut and coconut processing industries, as well as the empowerment of coconut planters through a partnership management-based economic togetherness system with institutional strengthening of farmers through community economic institutions.

4. Conclusion

Value chain approach in coconut farming system, synonymous with sustainable business development for farming communities in rural areas in the future, comes as a major actor. The implementation of the value chain provides added value in a sustainable manner and the creation of harmonious communication between coconut business actors. Value chain approach in coconut farming is still likely to be improved 3–10 times from current conditions.

The development of the maximum added value of coconut farming requires collective action through the sale of products, the purchase of production facilities, investment capital, as well as access to information about new technologies. Success in the development of coconut farming, the most decisive being human factor in this regard, concerns the attitude of the main actors, especially if the farming community itself is supported by related supporting elements. At the initial stage, the attitude of farmers' institutional managers is decisive for the achievement of success. The main thing that must be instilled in the institutional managers of agricultural communities is to respond positively to this institution as the main foundation for realizing the welfare of agricultural communities in the future. It is hoped that this institution will become an economic and social force in the village that can access the needs of farmers in the management of coconut farming.

The future of farming communities in rural areas is expected to contribute to not only the economic growth of every farmer and region but also more than that. It is expected that there will be a measurable increase in welfare, realizing the guarantee of a decent farming family life on an ongoing basis. Using the value chain analysis approach, it is expected that the farming community will gain insight and benefit from the program. The goal is for each member of the farmer group and combined farmer group to earn a minimum income of Rp 50 million/ha/year in the next 5 years. Subsequently, the development plan will be updated every 5 years, with the target of increasing the production of goods by at least Rp 50 million/ha/year. The principle of regional autonomy should ensure the economic well-being of each village, where all the assets of the farming community should be utilized to achieve yearly results.

Author details

Agustinus N. Kairupan[1*], Jantje G. Kindangen[2], Gabriel H. Joseph[3], Ronald T.P. Hutapea[2], Ibrahim Erik Malia[4], Paulus C. Paat[1], Derek Polakitan[1], August Polakitan[5], Jefny B. Markus Rawung[4], Meivie Lintang[3], Joula O.M. Sondakh[2], Payung Layuk[3], Olvie Grietjie Tandi[4], Herlina N. Salamba[4], Nelson H. Kario[2], Jonathan Anugrah Lase[6] and Rindengan Barlina[7]

1 Research Center for Animal Husbandry, National Research and Innovation Agency, Manado, Indonesia

2 Research Center for Macroeconomic and Finance, National Research and Innovation Agency, Manado, Indonesia

3 Research Center for Agroindustry, National Research and Innovation Agency, Manado, Indonesia

4 Research Center for Horticultural and Estate Crops, National Research and Innovation Agency, Manado, Indonesia

5 Research Center for Food Crops, National Research and Innovation Agency, Manado, Indonesia

6 Research Center for Animal Husbandry, National Research and Innovation Agency, Cibinong-Bogor, Indonesia

7 Research Center for Food Technology and Processing, National Research and Innovation Agency, Manado, Indonesia

*Address all correspondence to: audikairupan@gmail.com

References

[1] Statistical Yearbook of Indonesia 2022. Badan Pusat Statistik/BPS-Statistics Indonesia.

[2] Statistical of National Leading Estate Crops Commodity 2019-2021. Jakarta: Secretariat of Directorate General of Estate Crops; 2020

[3] Allorerung D, Advantages HT, Effendi DS, Maya IN, You CT, Supriadi H, et al. Executive Summary of Research Results in Collaboration with Dikti, Private/Local Government/ Other Institutions, and International Institutions. Jakarta: Agency For Agricultural Research; 2003

[4] Hernando. Agricultural Science. Jakarta: Penebar Swadaya; 1998

[5] Kaplinsky R, Morris M. A Handbook for Value Chain Research. Brighton, United Kingdom: Institute of Development Studies, University of Sussex; 2001

[6] Poole ND. Value chain perspective and literature: A review. Food Chain. 2013;**3**(3):199-211

[7] Stringer R. Value Chain Analysis. In: Workshop Value Chain Analysis Tanggal 5-7 Juni 2009 di Mataram NTB. Jakarta: Badan Litbang Pertanian; 2009

[8] Howieson J, Lawley M, Hastings K. Value chain analysis: An iterative and relational approach for Agri-food chains. Supply Chain Management: An International Journal. 2016;**21**(3):352-362. DOI: 10.1108/SCM-06-2015-0220

[9] Kotler P, Keller KL. Marketing Management. 15th ed. Pearson Edition Limited; 2016. Essex, Harlow: Pearson Education Ltd; 2016

[10] Rara ATS. Become A Successful Manager. Alfabeta, Bandung: Through Four Aspects of The Company; 2009

[11] Pantoja S. Agribusiness system approach in agricultural development. People's Economic Journal Th. II. 2003;**4**

[12] Yusuf, Kindangen JG. Value chain initiation in the development of people's coconut agribusiness. In: Handbook of Measuring Coconut Farming Innovation. Jakarta: Publisher Agro Indo Mandiri; 2020

[13] Adha Panca Wardanu and Uliyanti. Added value analysis and feasibility study of Nata De coco agroindustry In Ketapang Districs West Borneo. Journal Teknologi dan Industri Pertanian Indonesia. 2015;**07**(2):44-47

[14] Silaban R, Manullang RS, Hutapea V. Manufacture of virgin coconut oil (VCO) through a combination of fermentation and enzymatic techniques using pineapple extract. Journal Of Chemical Education. 2014;**6**(1):91-100

[15] Afidah E, Prayitno RS, Rumiyadi. Analysis value-added of coconut water to *Nata De coco* (case study In home industry ³Nata Mekar′). Journal Agroteknologi Merdeka Pasuruan. 2022;**6**(1):1-8

[16] Sinambela PD, Wu S, Rorielohoo DH, Silangen. Development of Shell Charcoal Making in North Sulawesi. Manado: Institute of Industrial Research and development; 1983

[17] Said EG, Rachmayanti and M. Z. Muttaqin. Agribusiness Technology Management Is the Key to the Global Competitiveness of Agribusiness Products. Bogor: PT. Ghalia Indonesia with MMA IPB; 2001

[18] Awang SA. Kelapa, Kajian Sosial Ekonomi. Yogyakarta: Aditya Media; 1991

[19] Susilowati R, Afiza Y. Business analysis and added value of coconut Shell charcoal In Tempuling District, Indragiri Hilir regency. Journal of Agribusiness Unisi. 2020;**9**(2):103-112

[20] Manikantan MR, Pandiselvam R, Beegum S, Mathew AC. Harvest and postharvest technology. In: The Coconut Palm (Cocos nucifera L.) - Research and Development Perspectives. Singapore: Springer; 2019. DOI: 10.1007/978-981-13-2754-4_13

[21] Prihanto H, Usmar. Company value factor analysis through earnings quality As an intervening. International Journal of Entrepreneurship. 2022;**26**(1):1-14

[22] Safitri Y, Abidin Z, Rosanti N. Performance and Added Value of CocoFiber Agroindustry In Kawasan Usaha Agroindustri Terpadu (KUAT) At Pesisir Selatan Sub-District of Pesisir Barat District. Vol. 2 No. 2. Lampung: JIIA; 2014

[23] Marsigit W. Development of Bengkulu local food processing products diversity to support sustainable food security. AGRITECH. 2010;**30**(4):256-264

[24] Mahmud Z. Crop the sidelines under the coconut. Journal of Agricultural Research and Development. 1998;**XVIII**(2):61-67

[25] Kindangen JG, Matindas LA, Layuk P, Raw JBM. Proposed Integrated Coconut Industry Development Policy Concept in Rural North Sulawesi. Manado: BPTP North Sulawesi; 2005

[26] Dornan J, dan JC. Maxwell. Strategy to Success. Your Step by Step Introduction to Success. USA: Network Twentyone; 1998

[27] Rintuh C, Miar. Institutional and people's Economy. first ed. Yogyakarta: BPTE; 2005

[28] Law of the Republic of Indonesia Number 32 of 2004. Article 213, On Regional Autonomy; 2004

[29] Nasution A, Rachmat M. Smallholder coconut agribusiness in Indonesia: Constraints and prospects. Agro Economic Research Forum. 1993;**10**(2-1):19-28

[30] Pratomo MN. The selling price of North Sulawesi copra is improving. Bisnis.com. 2020. Available from: https://Sulawesi.bisnis.com/read/20200107/540/1187665/earlyl-2020-price-selling-copra-sulut-improved

[31] Depends. Plantation office of North Sulawesi, encourage farmers by coconut oil. Between News. 2020. Available from: https://www.antaranews.com/news/1276101/Department of plantation-North Sulawesi-push-farmers if coconut oil

[32] Kairupan AN, Joseph GH, Kindangen JG, Paat PC, Polakitan A, Polakitan D, et al. Local tubers production and value chain: Evidence from Sangihe Island regency, Indonesia. In: Alem H, Jena PR, editors. Handbook of Sustainable Agricultural Value Chain. London: Intechopen; 2022. pp. 69-88. DOI: 10.5772/intechopen.103711

Chapter 8

Are There Adequate Incentives for Research and Innovation in the Plant Breeding Supply Chain?

Predrag Rajsic, Richard Gray, Alfons Weersink and Istvan Rajcan

Abstract

The breeding supply chain has expanded with genomic technology to include basic research scientists and applied genomicists along with traditional plant breeders and farmers. Genomicists have focused on identifying specific DNA sequences or quantitative trait loci (QTL) that can be used as molecular markers. However, the use of molecular marker-assisted selection (MAS) by breeders in their programs requires the identified QTL to be reliably correlated with agronomically desirable traits. Replication research is critical for reducing the risk associated with the adoption of new marker-based (or QTL-based) selection strategies, but the applied scientists doing genomics research often do not have the incentives to do replication and other research required to verify the reliability of markers. The misalignment of incentives in the breeding supply chain can curtail the development of the projected advances in food production by genomics research. Using a sample of 24 genomic journals, we found more highly ranked journals tend to favor new research on identifying new QTL over replication research on previously identified QTL. Given that breeders will tend to adopt only those markers perceived to be reliable, the implicit lack of incentives for basic and applied genomic scientists to undertake replication research can impede agricultural innovation.

Keywords: genomic technology, plant breeding, replication research, QTL discovery, supply chain

1. Introduction

The challenge of meeting the food and energy needs of a growing, increasingly wealthy, and global population from a finite and increasingly compromised resource base is formidable [1]. The hope of meeting the challenge rests with continued scientific progress and innovation [2]. Technological developments through the 20th century, largely fueled by publicly funded research programs, allowed the amount of agricultural output to nearly quadruple while the weighted index of real prices for 18 associated products has fallen by 75% [3]. Given the limited availability of untilled farmland and increasing agri-environmental constraints, future expansion will have to occur from increases in the productivity of cultivated land [4]. However, the rate of yield growth has generally declined since 2000 as governments have shifted funding to other priorities [5].

Scientific discoveries in the field of genomics create the potential for substantial future yield increases. The scope for innovation associated with genomics and related fields has some dubbing the 21st century as “the century of biology” [6]. In agriculture, genomic science is rapidly expanding the pool of knowledge that can accelerate the development of improved crops and animals. To realize this potential, it is important to consider the process of innovation and, in particular, how scientific knowledge gets translated into new products and processes.

Genomic research has resulted in the sequencing and mapping of the DNA for most large commercial crops and most of the important livestock species [7]. To date, genomicists, working with breeders, have been able to identify thousands of specific DNA sequences or quantitative trait loci (QTL) as molecular markers associated with many important (phenotypic) crop traits. As a result of these discoveries, many breeders are now using molecular marker-assisted selection[1] (MAS) to augment phenotypic selection in their breeding programs. Importantly, this knowledge is only useful to breeders when the QTL are reliable (i.e., they are highly correlated with the trait), the trait is commercially (agronomically) important, and MAS is cost-effective to deploy. The process of successful innovation or deployment of genomic knowledge requires the coordination and the activity of basic research scientists and applied scientists working with breeders and seed firms, the commercial development of sequencing platforms and equipment, and in the end, the farmer adoption of the new genetics in the form of new crop variety or animal genetics.

The breeding processes involving genomic research can be considered within a supply chain framework involving four relatively distinct groups: (1) basic research scientists who make discoveries useful to applied genomicists, (2) applied genomicists who develop molecular markers by identifying the associations between genes and the expression of a given trait (3) breeders who use MAS in their breeding program, and (4) farmers who incorporate the new varieties into their operation to increase production [8]. For this supply chain to operate effectively, each link must be strong and securely connected to the adjoining links in the chain. Each link must have the required resources and incentives to produce what is needed for uptake downstream.

This brings us to the issue addressed in this paper. While breeders need reliable markers (QTL), the applied scientists doing genomics research often do not have the incentives to do replication and other research required to verify the reliability of markers. The misalignment of incentives in the breeding supply chain has the potential to curtail the development of the projected advances in food production made possible through genomics research.

In this paper, we consider how the development and use of MAS in plant breeding can be influenced by the metrics used to reward basic and applied genomic scientists. We show that scientists will be incentivized to focus on new QTL discoveries at the expense of verifying the previously discovered QTL through replication research if the metric used to reward these scientists is the relative rank of the scientific journal in which their article appears. Journals are ranked by the number of times a scientific article is cited by later scientific articles, and it is more prestigious for a scientist to publish in higher-ranked journals. Any bias in the type of articles accepted by more highly ranked journals would

---

[1] MAS is a form of genomic selection where a relatively small number of genetic markers are used in the selection process.

affect the type of research carried out. We will show that more highly ranked journals publish more articles on MAS discovery rather than verification. Given that breeders will tend to adopt only those markers perceived to be reliable, the implicit lack of academic reward for verification studies can distort incentives for basic and applied genomic scientists to undertake replication research, which, in turn, can impede agricultural innovation. Having highlighted this issue, we make a more general argument that applied genomic researchers have a unique role in innovation systems and, therefore, require a modified set of incentives rather than just relying on journal ranking alone.

The replicability crisis identified by Ioannidis [9], later documented by Open Science Collaboration [10], Baker [11], and Fanelli [12] compounds the incentive compatibility problem. Ioannidis [9] finds that most biometric studies may be false or an expression of a prevailing bias. Open Science Collaboration [10] conducted replications of 100 experiments published in three high-ranking psychology journals in 2008 and found that while 97% of the original experiments had statistically significant results only 36% of the replications had statistically significant results. In a survey of scientists' views of reproducibility by Baker [11], over 70% of the 1500 researchers surveyed reported that they have tried and failed to reproduce another scientist's experiments and 52%, 38%, and 7% of respondents stated that there is a "significant reproducibility crisis," "slight crisis," or "no crisis," respectively. Fanelli [12] concludes that although science may not be facing a reproducibility crisis, reproducibility is an important challenge that needs to be addressed. All these issues create further disincentives for researchers attempting replication research and have subsequent implications for the design of research funding programs.

The private rewards for discovery research, including prestige from publication in a higher quality journal, should be higher given the higher costs and skills to undertake such research. However, the idea of incentive misalignment stems from the public aspect of the benefits of replication research along the supply chain. Replication research and new research can be viewed as complementary inputs in productivity growth stemming from plant breeding. Thus, the public marginal benefits of some new/discovery research may be wasted if not coupled with replication research. Our research does not prove that the mix of replication and new research is suboptimal, but it does give some evidence that it might be. Further research is needed to assess potential efficiency gains from alternative mixes of new research and replication research.

The remainder of the paper is organized as follows. We begin with a conceptual framework that is informed by a description of the research and development required for the implementation of MAS in plant breeding. We also describe factors that breeders weigh in their adoption decision. Given the literature on supply chain management (SCM) and incentive alignment, we review previous studies that have examined agricultural innovation systems and incentives. Based on this literature and the description of the research and development required for the implementation of MAS in plant breeding, we develop a potential rationale that suggests that there may be a lack of incentives for replication research in genomics. To assess this rationale, we examine the statistical relationship between marker discovery versus verification and journal ranking. Building on the results of this analysis, we show that output metrics based solely on journal ranking, create a strong incentive for genomic scientists to focus on discovery rather than verification, and without verification, the MAS is less likely to be adopted. We conclude the paper with a discussion of the implications, arguing that funding bodies and public administrators interested in innovation should consider a broader range of metrics regarding replication research, particularly given the value of applied research to downstream users.

2. Conceptual framework

2.1 The plant breeding supply chain and genomic selection

The use of marker-assisted selection (MAS) has become an important tool for the breeders of many crops. Recent advances in genomics have linked particular DNA molecular markers to the phenotype or function of the plant. Breeders are increasingly using these molecular markers to help, select, and screen the plants for their breeding program. For example, if a wheat breeder is interested in developing varieties resistant to a particular strain of leaf rust, the molecular marker for rust resistance can be used as a screen to identify lines with this resistance trait, saving time, and, potentially, cost in the breeding process. This usefulness of MAS has created a demand for applied genomics science where researchers work to find molecular DNA markers associated with important phenotypic traits in crops.

The use of MAS in breeding programs relies on the existence of markers of interest, their reliability, and the breeders' cost of implementation. For many crops, there is already a long list of molecular markers, often numbering in the thousands [13]. A breeder must judiciously decide which markers to use in their breeding program because the total number of markers that can effectively be used for variable selection is limited by the effect on plant populations. If the traits exist in a single gene of a heterozygous parent, only half of their offspring will inherit the trait. As such, every time a marker is used for selection, it will reduce the size of eligible lines in the first generation (F1) by 50%. If a breeder uses four markers to select the first generation, only one in 16 of the F1 generation would have all four markers. If 10 markers were used, 1 in 1024 would meet the screen, eliminating 99.9% of the lines. As such, breeders must very carefully consider which markers and how many markers they are going to use in their breeding program.

The cost of implementing MAS will also affect both the number of markers and the stage of implementation in the breeding program [13]. One of the strongest drivers of increased MAS adoption has been the decrease in the per-sample cost of analysis using technology. Early technologies had limited scope and were time-consuming and labor-intensive, resulting in large costs per sample. When MAS application was expensive, breeders would either avoid using them, use them for the parent lines only, or use them later in the breeding process at F4 or F5 generation, after they had considerably narrowed the potential number of lines by other means. As the sample cost has fallen, more breeders are able to use MAS in the selection of F1 generation.

Marker reliability is critically important for breeder adoption. Markers are typically discovered through a statistical association of a particular phenotype and the genetic marker. If the marker and a single gene responsible for the trait are closely located on the same chromosome, there will be a high correlation between the existence of the marker and the phenotype produced by that gene. If, however, the marker is not close to the gene responsible for the phenotypic trait, the gene can exist in the absence of the marker (i.e., a type 1 error) and the marker can be present without the phenotypic trait (i.e., a type 2 error). A marker is considered reliable when both type 1 and type 2 errors are close to or equal to zero.

The economic costs of both type 1 and type 2 errors are important. In the case of type 1 error, selecting a line that does not have the desired trait increases the cost of all subsequent downstream development, until the error is found. For example, if a program carries 20% of lines that do not have the desired trait, the costs per viable line would increase proportionally. With type 2 errors, rejecting a line that in fact does have the desired trait reduces the number of lines that could be used for subsequent selection. A 15% chance of a type 2 error at the F1 stage would require the breeder to have a population 15% larger to yield the same number of viable lines after the MAS. Given the economic importance of marker reliability, it is easy to understand why breeders often identify marker reliability as a key factor in the adoption decision.

The reliability of markers can be improved by further genomic research that tests the relationship between MAS and desired genes in a larger population while looking for more reliable markers at each stage. The actual gene (which is the perfect marker) can be identified often through a process of elimination and more recently can be verified through gene editing techniques. Given both the theoretical and empirical evidence pointing to the importance of marker reliability, research that verifies and improves marker reliability is valuable for downstream breeders and innovation outcomes.

The development and use of MAS is an excellent example of how certain advances in science can be mobilized for breeding, resulting in improved crop varieties and improved agricultural productivity. If one conceptualizes this as an innovation system involving components of basic science, genomics, applied science, breeder, and commercial use, many individuals and institutions are involved.

2.2 Agricultural innovation systems and incentives

Well-functioning supply chains require communication, coordination, and an alignment of incentives for participants along the chain. While markets can play an important role in the coordination along a supply chain, markets also often fail, requiring other mechanisms including contracting or other forms of vertical integration to provide effective low-cost coordination. Firms or industries that are successful in developing cost-effective supply chains gain a competitive advantage and can prosper, while those that fail to efficiently achieve the coordination required, lose market share and can be driven out of the market. Porter [14] argues that communication and knowledge flows are key to innovation and the competitive advantage of nations.

An important branch of the SCM literature examines innovation systems that allow firms, industries, and countries to create and mobilize knowledge to increase productivity [15]. In our experience, agricultural researchdiffers from many industries because of the extensive involvement of the public sector throughout the breeding supply chain from basic and applied research to crop breeding and agronomy to farm extension programs. However, many of the public-based systems have evolved to include private research and commercialization firms within the supply chain. A wide variety of institutional arrangements involving public-private partnerships exist in crop innovation systems globally [16]. In the case of MAS, much of the basic science, marker discovery, and verification continues to be done primarily by the public sector, with private firms increasingly engaged in breeding and variety development.

There are many examples of highly successful crop innovation systems that have linked cutting-edge genomics with the widespread deployment of MAS and other applied tools. In the United States, the United States Department of Agriculture (USDA) and other public scientists have undertaken the basic research and genomics that have supported the development of very large and sophisticated private molecular breeding programs in soybeans and corn) [17]. In Canada, Genome Canada has supported many large crop genomics projects that are integrated into public, producer, or privately supported breeding programs. In France, BreedWheat, a large public-private research consortium made up of 14 organizations, has been collaborating since 2010 to undertake genomics research with the goal of supporting the development of wheat varieties [18].

There are also many examples where effective supply chains for MAS have failed to develop [19]. For example, in our own soybean breeding and genetics lab at the University of Guelph, we have developed many markers through graduate student projects that were published but never used in MAS (Rajcan, personal communication). These failures can be roughly attributed to a lack of resources and/or incentives required to support the genomic research necessary to develop reliable molecular markers. Generally, these failures tend to be associated with either minor crops or crops primarily grown in countries with limited public resources [19]. Other failures in MAS adoption have occurred where the markers exist but breeders lack the knowledge and/or cost-effective tools to undertake and deploy MAS.

A lack of coordination can impair a supply chain even when adequate resources exist. One classic failure occurred in the United Kingdom after the privatization of Cambridge's highly effective Plant Breeding Institute. After the sale, genomic scientists that had been part of the institute were strongly incentivized by the research-granting councils to focus on research with scientific impact as measured through the rank of the scientific journal where the research is published [20]. With these incentives, most of the wheat genomic scientists shifted their focus away from wheat toward model organism crops such asArabidopsis that are genetically simpler and for which genomic analysis could be generated sooner and published in higher-ranking journals. This diversion of effort left the private wheat research industry without the support of public scientists for nearly 10 years. Recognizing the problem, the principal funder of basic and strategic biological research in the United Kingdom, the Biotechnology and Biological Sciences Research Council (BBSRC), along with other research granting councils, created new programs and incentives for the basic scientist to work with the private breeders, largely resolving the coordination issue [20].

2.3 Replication research in the breeding research and innovation process

The breeding process occurs through a functional interaction of four relatively distinct groups: basic research scientists, applied genomicists, breeders, and farmers. Depending on the institutional (ownership) structure, these interactions can occur within and across at least two dimensions: private and public [21]. Each of the distinct functional groups has a set of goals that can potentially be met by the outputs of other functional groups in the breeding process. These goals shape the incentives for choices of production activities among the groups. The result of the activities of all four groups is the production of agricultural crops that ultimately satisfy consumer demand.

Although the purpose of the plant breeding research and innovation process can be defined as serving the end-users of the crops, each group within the process may have its own immediate goals, which are shaped by the formal or informal metrics used to measure success within each group. The most commonly used metric for determining the performance of academic researchers who tend to focus on basic research rather than on applied research is the number of peer-reviewed publications in top journals and the number of citations of their published work [22, 23]. In the context of gene mapping, this has generally meant publishing discoveries of new genotype–phenotype associations (QTL), with less interest within the academic community in the subsequent replication and verification of QTL [24–27]. This indicates that the publication of new QTL discoveries represents a key measure of the academic genomicists' performance.

In our view, rather than discovering new QTLQTL, the focus of applied genomicists is primarily on the verification of QTLQTL that can be used for commercial purposes. Only QTLQTL with a certain degree of stability over populations and environments have the potential to be commercially viable. Thus, in addition to the incentives for discovering new QTLQTL, private applied genomicists have incentives to verify the stability of newly discovered QTLQTL. The set of these QTLQTL is established by the academic genomicists who may be seeking QTLQTL that lead to quick publication rather than those associated with commercially desired traits.

Breeders are interested in using the newly discovered QTL for selecting improved breeds and varieties. The goal is to select superior parent lines to create a population with distinct genetic features, which are associated with preferred phenotypic features. This distinct population can then be registered as a new variety or breed.

While profit is the key performance measure for a private breeding program as a whole, the number of new licensed varieties is the main performance measure for public plant breeders. Both private and public plant breeders have incentives to use the results of published QTL studies if they expect these results to be effective means of developing a new variety or a breed. The additional criterion for choosing which varieties will be developed by private breeders is the profitability of the new variety. In cases where the QTL discoveries, reported in peer-reviewed journals, are not sufficiently replicated, breeders need to determine which QTL would be useful for each population anew. Due to the sheer volume of academic QTL publications, the costs of determining which QTL could potentially be useful for private breeders may be high. All this increases the costs to plant breeders of adopting MAS strategies compared to a situation when genotype–phenotype relationships are validated through verification studies.

The costs for plant breeders to adopt MAS may lead to low incentives for using genetic markers as a means of developing new varieties or breeds and thus potentially lower the probability of developing a new variety or breed. Collaboration between geneticists and breeders remains an important challenge [28, 29]. Xu and Crouch state that [30] "high proportion of published markers [are] failing at one or more of the translation steps from research arena to application domain." Similarly, hold that "MAS has had only a small impact on plant breeding so far." As one of the reasons for this low impact, they identify the low publishing potential of QTL validation studies:

> *"New QTL are frequently reported in scientific journals, but reconfirmation of these QTL in other germplasm and identification of more useful markers are usually not considered novel enough to warrant new publications. This is unfortunate because it is exactly this type of information that is needed for MAS."*

Ref. [31] also note that the "vast majority of publications on the subject are not considered to have real impact on breeding efforts."

Thus, the public applied genomicists' incentives to publish new research at the expense of replication or verification studies may not be in alignment with the breeders' needs for reliable genetic information. Note that although MAS has been adopted in breeding programs for several major crop species, the "majority of the legume crops ... remained untouched with genomics revolution." Similarly [32], hypothesize that "top journals, concerned with the need to maintain reputations and encourage originality, may be less likely to publish replications." To assess this hypothesis, we measured the extent of replication research in a sample of genetics journals. We also tested the link between journal rank and the tendency to publish discovery versus replication research.

3. Empirical model

Our investigation of the extent of incentive incompatibility within the plant breeding supply chain consists of two main steps. First is the gathering of data on scientific articles on QTL discovery research and QTL replication research along with the ranking of the journals in which they appeared. Second is the determination of the relationship between the type of QTL research and the journal ranking to test the hypothesis that more prestigious journals publish more articles on MAS discovery rather than verification. The next section of the paper discusses the implication of the results on the design of research programs to ensure incentive compatibility.

Keywords were identified that could be used to search for QTL discovery or replication research in Google scholar. The preliminary search criterion was initially selected through consultations with plant breeding experts. The selection process was supplemented by determining the frequency of keywords and the context in which they were used for the 2019 volume of four genomic journals: Theoretical and Applied Genetics, Euphytica, Journal of Genetics, Genetic Resources and Crop Evolution.

The results from the search of all papers published in 2019 of the four journals are summarized by journal and type of research in **Table 1**.[2] Papers using phrases "QTL Discovery," "New QTL", "Novel QTL," or "Identified QTL" tended to be the most numerous in theoretical and applied genetics and euphityca. Papers using these phrases also tended to be focusing on the identification of particular QTL rather than on general method development for QTL identification. On the other hand, when associated with the words "marker," "gene or allele," the attributes "new," "novel," "identified," and "developed" tended to appear in articles that are focusing on developing methods rather than on the identification of particular QTL.[3] When it comes to replication research, "verification," "replication," or "confirmation" of a QTL were the phrases most frequently associated with replicating previous QTL research. Based on these findings, we selected "new QTL," "novel QTL," and "QTL discovery" as indicators of new QTL research, and "validated QTL," "confirmed QTL," "verified QTL," "QTL validation," "QTL verification," and "QTL confirmation" as indicators of replication research.

---

[2] We have prepared a summary of key points for each article that contains keywords denoting new QTL research or replication research. This document is available as supplementary material. **Table 1** is the condensed summary of this document.

[3] When "new," "novel," "identified," and "developed" were associated with markers, genes or allele they focused on the identification of markers, gene or alleles that will help discover future QTL. They would not necessarily outline a complete method to discover QTL but would rather state that this marker will help to discover novel QTL responsible for a certain trait.

		Journal[a]				
Research Type	**Keywords used to classify articles**	**TAG**	**EUPH**	**GRCA**	**JG**	**Primary research focus**
New	Discovery/New/Novel/Identified QTL	64	37	2	2	Identifying particular QTL
	Discovery/New/Novel/Identified/ Developed Marker	14	7	15	10	Developing methods
	Novel/Identified gene	20	14	3	4	Developing methods
	Novel/Identified allele	3		6	1	Developing methods
Replication	Verification/Replication/Confirmation of a QTL	22	14	0	2	Identifying particular QTL
	Verification/Replication/Confirmation of a gene	2	7	0	0	Developing methods
	Verification/Replication/Confirmation of marker	0	3	0	3	Developing methods

[a]*Journals are Theoretical and Applied Genetics (TAG), Euphytica (EUPH), Genetic Resources and Crop Evolution (GRCE), and Journal of Genetics (JG).*

Table 1.
Frequency of keywords by QTL research type and context for four academic journals in 2019.

A variable was created for each journal to proxy the relative importance of discovery QTL research to overall QTL research published in that journal. Four ratios were calculated to determine the relative importance of discovery versus replication research for each journal. In addition, a fifth ratio was developed to measure the importance of overall QTL research (new or replication-related) in the selected journals. The first three ratios focus on individual keywords associated with new research while the fourth ratio puts all three keywords together. This approach allows us to identify keywords that may be more relevant than others as indicators of differences across journals.

The first ratio, *Share* New_i, is a measure of the importance of new QTL research relative to overall QTL research for journal i.

$$ShareNew_i = \frac{New_i}{New_i + Replication_i} \tag{1}$$

where New_i represent the number of papers in which "new QTL" appeared in journal *i's* online search form[4], and $Replication_i$ represents the total number of papers in which "verified QTL," "confirmed QTL," or "Validated QTL," or "QTL verification," or "QTL confirmation," or "QTL validation" appeared in journal i's online

[4] The journals' online search pages did not have a specified date range. The results were from any volume/ issue of the journal that would be available online and on the journal's website.

search form. A second alternative uses the keyword “QTL discovery” to define the following ratio

$$ShareDiscovery_i = \frac{Discovery_i}{Discovery_i + Replication_i} \quad (2)$$

where $Discovery_i$ represents the number of papers in which “QTL discovery” appeared in journal *i’s* online search form. A third uses the keyword novel to represent new QTL research,

$$ShareNovel_i = \frac{Novel_i}{Novel_i + Replication_i} \quad (3)$$

where $Novel_i$ represents the number of papers in which “Novel QTL” appeared in journal *i*’s online search form. Finally, all three keywords were used in the following ratio

$$ShareNewQTL_i = \frac{New_i + Discovery_i + Novel_i}{New_i + Discovery_i + Novel_i + Replication_i} \quad (4)$$

The numerator is the number of papers in which all three keywords indicating new QTL research appeared in a given journal, while the denominator is the number of papers in which all the selected keywords appeared in the same journal. In all four ratios, the higher the value, the greater the importance of marker discovery in the publication of QTL research in that journal. It is important to note that the maximum value of any given ratio is one, which implies that all keywords are associated with new/discovery/novel QTL research. The fifth ratio, $ShareQTL_i$ represents the share of papers in which any of the selected keywords (both new and replication-related) appeared in the total number of papers published in journal i:

$$ShareQTL_i = \frac{New_i + Discovery_i + Novel_i + Replication_i}{AllPapers_i} \quad (5)$$

The purpose of this ratio was to assess the overall importance of QTL research across the 24 selected journals and to test for differences in the ratio between higher-ranked journals and lower-ranked journals.

The five ratios defined in Eqs. (1)–(5) were calculated from any volume/issue of a specific journal that was available online and on the journal’s website for 24 journals selected in the crop genomics discipline. The ranking of the 24 journals according to their respective average scientific journal rankings (SJR) indicators since 2001 as reported by the SJR (2019) database is listed in **Table 2**. The SJR indicator weighs the influence of a journal, based on the number of citations received by the journal and importance of the journals from which the citations are derived. The higher the SJR indicator, the greater is the prestige of the journal.

The first step in the analysis is to examine the annual frequency of the three new QTL keywords (new, discovery, and novel), the frequency of the six replication QTL keywords (validated, confirmed, verified, validation, verification, and confirmation), and the resulting share measures given by Eqs. (1)–(5). Using the four keywords for QTL discovery research and the five keywords for QTL verification research, Google Scholar was searched to identify refereed articles containing those terms in academic

Rank	Journal	SJR indicator
1	Trends in Genetics	7.49
2	Current Opinion in Plant Biology	5.49
3	PLoS Pathogens	4.56
4	Plant Physiology	3.77
5	Genetics	3.60
6	Evolution	3.39
7	Molecular Ecology	3.15
8	Journal of Experimental Botany	2.26
9	Journal of Evolutionary Biology	2.21
10	BMC genomics	2.00
11	Theoretical and Applied Genetics	2.00
12	Genetics Selection Evolution	1.19
13	Molecular Breeding	1.18
14	BMC Genetics	1.08
15	Journal of Heredity	1.08
16	Crop Science	1.00
17	Cytogenetic and Genome Research	0.95
18	Genome	0.94
19	Tree Genetics & Genomes	0.91
20	Journal of the American Society for Horticultural Science	0.76
21	Euphytica	0.72
22	Journal of Animal Breeding and Genetics	0.69
23	Genetic Resources and Crop Evolution	0.61
24	Journal of Genetics	0.36

Table 2.
Journals used for calculating the relative frequency of keywords referring to new and replication research in genetics and their respective SJR score between 2001 and 2018.

journals between 2000 and 2019.[5] The average of the numbers of papers containing the respective keywords and the share measures are compared to determine if there is a statistically significant difference between the top 12 ranked journals from the other 12 with lower SJR rankings. In addition, trends are examined to determine if the prominence of the keywords has changed over time.

Finally, a regression analysis is conducted with the five share measures for each journal regressed against its SJR indicator value.

$$Share\, J_i = \beta_0 + \beta_1\, SJR_i + e\, J = New, Discovery, Novel, New\ QTL, and\ QTL \tag{6}$$

[5] This refers to the presence of said phrases anywhere in the paper, not to the keywords listed at the beginning of a paper.

where β_0 and β_1 are parameters to be estimated. The implied hypothesis is that journal indicators may influence the motivation of the journal to publish new research at the expense of replication research. Alternatively, the causality may run in the opposite direction. It may be that the relative share of new research boosts the ranking of a journal. Either way, there would be a relationship between the quality of the journal and the relative share of new research, and the main aim of our research is to investigate this potential relationship. We hypothesize that a higher SJR indicator is associated with a propensity to publish a greater share of new QTL research, which means that journals with higher SJR indicators would also tend to have higher values for the ratios in Eqs. (1)–(5).

To truly test the claim that top journals have a motivation to publish new research at the expense of replication research, one would need to analyze the rejection rates for new research versus replication research rather than the number of observed published articles. Unfortunately, the data on rejection rates are not available. This is our best attempt to get around this data problem. The downside is that the strength of our conclusions is reduced. Thus, rather than testing the hypothesis of journal bias in favor of new research over replication research, we are assessing potential links and relationships.

4. Results

The number of papers containing the selected keywords related to QTL research that are available in the Google Scholar database for all journals from the years 2000 through 2019 are listed in **Table 3**. The frequency of papers with QTL described as *New* or *Discovery* or *Novel* has increased steadily since 2000. The trend variables suggest that each of the research articles containing the key search words "*New*" and "*Discovery*" have increased annually by approximately 11 papers per year while the number of papers with "*Novel*" has increased by nearly 17 papers per year. Articles with keywords for replication research have also increased over time, but the absolute rate of annual increase is significantly less than the increase in papers containing new QTL keywords.

	Papers with new QTL keywords			Papers with replication QTL keywords		New QTL as share of total measures			
Year	**New**	**Discovery**	**Novel**	**Verified or confirmed or validated**	**Verification or confirmation or validation**	**Share new**	**Share discovery**	**Share novel**	**Share new QTL**
2000	40	4	14	5	7	0.77	0.25	0.54	0.83
2001	42	5	8	8	10	0.70	0.22	0.31	0.75
2002	51	12	17	8	29	0.58	0.24	0.31	0.68
2003	68	19	23	14	37	0.57	0.27	0.31	0.68
2004	92	42	38	18	33	0.64	0.45	0.43	0.77
2005	91	43	53	29	37	0.58	0.39	0.45	0.74
2006	135	34	70	13	52	0.68	0.34	0.52	0.79
2007	119	57	69	18	61	0.60	0.42	0.47	0.76
2008	131	60	76	29	73	0.56	0.37	0.43	0.72

	Papers with new QTL keywords			Papers with replication QTL keywords		New QTL as share of total measures			
Year	New	Discovery	Novel	Verified or confirmed or validated	Verification or confirmation or validation	Share new	Share discovery	Share novel	Share new QTL
2009	158	83	94	41	83	0.56	0.40	0.43	0.73
2010	135	78	106	21	104	0.52	0.38	0.46	0.72
2011	159	89	121	34	111	0.52	0.38	0.45	0.72
2012	198	118	126	35	119	0.56	0.43	0.45	0.74
2013	203	108	139	37	92	0.61	0.46	0.52	0.78
2014	199	114	184	41	120	0.55	0.41	0.53	0.76
2015	211	128	187	38	138	0.55	0.42	0.52	0.75
2016	218	137	226	34	117	0.59	0.48	0.60	0.79
2017	204	174	265	41	129	0.55	0.51	0.61	0.79
2018	226	208	303	39	128	0.58	0.55	0.64	0.82
2019	257	220	403	38	133	0.60	0.56	0.70	0.84
Mean	146.85	86.65	126.1	27.05	80.65	0.59	0.40	0.48	0.76
Trend	11.06*	10.53*	16.92*	1.86*	7.18*	−0.006	0.014*	0.014*	0.003
Std Error	0.54	0.63	1.49	0.24	0.46	0.002	0.001	0.002	0.001

**Coefficient on trend variable is statistically significant at the 99% confidence level.*

Table 3.
Frequency of QTL research papers Google Scholar from 2000 to 2019.

Although the rate of increase is higher for the keywords related to new QTL discovery compared to the replication QTL keywords, the relative focus on new QTL research compared to replication research depends on the choice of keywords. The *Share New* and *Share New QTL* ratios did not change significantly over time (**Table 3**). However, the value of the *Share Discovery* and *Share Novel* increased by approximately 1.4 percentage points annually over the last two decades.

In order to assess whether the increasing focus on new QTL research versus replication has been driven in part by incentives for academic genomicists, we examined the frequency of the keywords for each of the 24 journals in **Table 4**. The selected journals in **Table 4** are listed in the order of their SJR ranking (see **Table 2**) and the number of articles, in which the keywords associated with new QTL research and replication QTL research, are given along with the four ratios indicating the share of new QTL research to total QTL research. The averages of these measures are calculated for the top 12 ranked and the bottom 12 of the journals chosen for analysis and a t-test[6] used to determine if the difference in averages is statistically significant.

The top 12 journals tended to have higher values for *"New," "Discovery,"* and *"Novel"* as hypothesized as compared to the bottom 12 of the journals selected. There is no difference in the average appearance of keywords related to replication between

[6] The test was paired, two samples for means.

Journal	New QTL keywords			Replication QTL keywords		Total publications	New QTL as share of total measures				Share QTL
	New	Discovery	Novel	Verified or confirmed or validated	Verification or confirmation or validation		Share new	Share discovery	Share novel	Share new QTL	
Trends in Genetics	49	6	1	0	0	6650	1.00	1.00	1.00	1.00	0.0084
Current Opinion in Plant Biology	2	5	4	0	3	2240	0.40	0.63	0.57	0.79	0.0063
PLoS Pathogens	2	1	3	0	1	3180	0.67	0.50	0.75	0.86	0.0022
Plant Physiology	7	10	7	2	7	102,000	0.44	0.53	0.44	0.73	0.0003
Genetics	79	23	23	21	23	121,000	0.64	0.34	0.34	0.74	0.0014
Evolution	5	1	1	0	0	102,000	1.00	1.00	1.00	1.00	0.0001
Molecular Ecology	6	6	5	0	2	15,100	0.75	0.75	0.71	0.89	0.0013
Journal of Experimental Botany	29	14	37	5	27	16,100	0.48	0.30	0.54	0.71	0.0070
Journal of Evolutionary Biology	0	1	0	0	0	4640	0.00	1.00	0.00	1.00	0.0002
BMC Genomics	36	23	34	13	16	14,500	0.55	0.44	0.54	0.76	0.0084
Theoretical and Applied Genetics	177	54	129	83	100	12,700	0.49	0.23	0.41	0.66	0.0428
Genetics Selection Evolution	21	6	15	7	2	1890	0.70	0.40	0.63	0.82	0.0270
Average for the top 12 journals based on SJR Indicator	34.42	12.50	21.58	10.92	15.08	33,500	0.59	0.59	0.58	0.83	0.0088
Molecular Breeding	70	34	56	43	60	5580	0.40	0.25	0.35	0.61	0.0471
BMC Genetics	23	8	19	14	8	4130	0.51	0.27	0.46	0.69	0.0174
Journal of Heredity	14	1	3	3	3	8150	0.70	0.14	0.33	0.75	0.0029
Crop Science	66	39	45	32	29	70,700	0.52	0.39	0.42	0.71	0.0030
Cytogenetic and Genome Research	2	0	0	1	0	6180	0.67	0.00	0.00	0.67	0.0005

DOI: http://dx.doi.org/10.5772/intechopen.110347

	New QTL keywords			Replication QTL keywords		Total publications	New QTL as share of total measures				Share QTL
Journal	New	Discovery	Novel	Verified or confirmed or validated	Verification or confirmation or validation		Share new	Share discovery	Share novel	Share new QTL	
Genome	9	3	13	0	0	5867	1.00	1.00	1.00	1.00	0.0043
Tree Genetics & Genomes	9	12	1	4	14	2080	0.33	0.40	0.05	0.55	0.0192
Journal of the American Society for Horticultural Science	1	2	4	1	1	4140	0.33	0.50	0.67	0.78	0.0022
Euphytica	82	16	43	29	31	9860	0.58	0.21	0.42	0.70	0.0204
Journal of Animal Breeding and Genetics	10	1	2	2	6	1860	0.56	0.11	0.20	0.62	0.0113
Genetic Resources and Crop Evolution	0	2	4	2	1	2860	0.00	0.40	0.57	0.67	0.0031
Journal of Genetics	6	0	4	1	1	81,900	0.75	0.00	0.67	0.83	0.0001
Average for the bottom 12 journals based on SJR Indicator	24.33	9.83	16.17	11.00	12.83	16,942	0.53	0.31	0.43	0.71	0.0110
difference in mean between top 12 and bottom 12	10.09	2.67	5.40	−0.08	2.25	16,557	0.06	0.28*	0.15*	0.12*	0.0022
t-test statistic	0.52	0.41	0.40	−0.01	0.20	1.15	0.69	3.22	1.54	2.65	

**Difference in averages is statistically significant at the 90% confidence level.*

Table 4.
Frequency of Papers with QTL research keywords in each of 24 journals by SJR ranking from 2000 to 2019.

the two groups of journals categorized by SJR ranking. Given the greater appearance of new QTL keywords in the top 12 ranked journals compared to the other 12 and the insignificant difference QTL replication keywords between the two groups of journals, the relative role of new QTL as a share of total QTL research is higher in the top-ranked journals is expected. The difference is particularly high for the *Share Discovery* ratio and statistically insignificant for *Share New*.

Next, the difference in the share of papers mentioning QTL research (new or replication) between the top 12 and the bottom 12 journals was not statistically significant. This indicates that there was no significant difference in the level of emphasis on overall QTL research between the two groups. This suggests that, even though both groups of journals publish QTL research, the top group puts more emphasis on new QTL research while the bottom group puts more focus on QTL replication research. The total number of papers published was higher in the top 12 journals. This might suggest that QTL research in the top journals competes with a greater number of topics than in the bottom 12 journals. Higher-ranked journals may publish cutting-edge lines of research that are not present in the lower-ranked papers. In this setting, replication QTL research may be at a competitive disadvantage when competing not only against new QTL research but also against other advanced lines of research in top-ranked journals.

A final step in the empirical analysis is to examine the relationship between a journal's SJR indicator and the value of the four ratios used to proxy the relative importance of new QTL research to overall QTL research in the journal. The results of the five regressions (Eq. 5) are listed in **Table 5**. There is a positive relationship between the focus on new QTL discovery and journal rank as defined by its SJR indicator as hypothesized for the first four ratios. The fifth ratio, measuring overall QTL research, had a negative but not statistically significant coefficient on Share QTL. The intercepts were positively significant at a 99% confidence level for all five models, which is expected as the SJR indicator is generally a number greater than zero for most journals. The slope for *Share Discovery* was positive and significant at the 99% confidence level, while the slopes for *Share Novel* and *Share New QTL* were positive and significant at the 95% confidence level. This result is consistent with the hypothesis

Dependent variable	Intercept	SJR indicator	R^2	Adjusted R^2
Share New	0.4721**	0.0416	0.0841	0.042
	(0.0812)[a]	(0.0416)		
Share Discovery	0.2461**	0.095**	0.308	0.276
	(0.0841)	(0.0304)		
Share Novel	0.3656**	0.0643*	0.168	0.130
	(0.0846)	(0.0305)		
Share New QTL	0.6991**	0.0344*	0.2245	0.189
	(0.0378)	(0.0136)		
Share QTL	0.013032** (0.0042)	−0.0015 (0.0015)	0.0411	−0.002

[a]*Standard errors are in parentheses.* ***Significant at the 99% confidence level.* **Significant at 95% confidence level.*

Table 5.
Regression (n = 24) results between focus on new discovery research and SJR indicator (Eq. 5).

that more highly ranked journals tend to favor new research over replication or verification research, while the extent of overall QTL research is not significantly affected by journal rank.

5. Conclusions

The purpose of this paper was to develop a better understanding of incentives for research and innovation within the plant breeding process. Advances in genomic technology have brought the potential for significant gains in agricultural productivity within a much shorter time frame than possible with traditional phenotypic breeding strategies. The breeding supply chain has expanded with genomic technology to include basic research scientists and applied genomicists along with traditional plant breeders and farmers. Capturing the gains made possible by genomic technology will require cooperation through the key stakeholders within this plant breeding supply chain.

Genomicists have focused on identifying specific DNA sequences or QTLs that can be used as molecular markers. However, the use of MAS by breeders in their programs requires the identified QTL to be reliably correlated with agronomically desired traits. Replication research is critical for reducing the risk associated with the adoption of new marker-based (or QTL-based) selection strategies, but the applied scientists doing genomics research often do not have the incentives to do replication and other research required to verify the reliability of markers. The misalignment of incentives in the breeding supply chain can curtail the development of the projected advances in food production by genomics research.

The metric used to reward basic and applied genomic scientists is the prestige or higher rank of the journal where their research is published, and this has created a bias toward identifying new markers rather than the verification of existing markers. Using a sample of 24 genomic journals, we found more highly ranked journals tend to favor new research on identifying new QTL over replication research on previously identified QTL. Given that, breeders will tend to adopt only those markers perceived to be reliable, the implicit lack of incentives for basic and applied genomic scientists to undertake replication research can impede agricultural innovation.

However, there may be other factors influencing academic geneticists' decision to perform replication research. Rajsic et al. [33] find that cost considerations are important in determining the sizes of training populations and the number of replications. QTL validation examines whether the same QTL appears when the genetic background is grown in other locations and/or years and whether its effect can still be detectable when introduced into a different genetic background. Lack of appropriate funds for public research may contribute to an overall lack of replication studies published by academic geneticists. Although the lack of replication done by academic geneticists could explain an overall low ratio of replication research to new research, it is hard to see why this would cause differences between top-ranked journals and lower-ranked journals. This suggests that incentives play a role in addition to other potential factors.

Policymakers designing breeding research and innovation programs must recognize the potential for misalignment of incentives within the supply chain. Rather than reward applied scientific researchers on the basis of publication surrounding new QTL in high-impact journals, funding agencies should create incentives for basic scientists to work with the breeders to focus on the identification and replication of traits desired at the farm level. Alternatively, funders of large research projects targeted toward a variety of development could require, and/or fund, additional verification studies for new QTL. In the absence of policy change, the lack of verification will continue to be an impediment to crop innovation.

Acknowledgements

Funding for the research was partially provided by the Applied Bean Genomics and Bioproducts Project, a multi-interinstitutional partnership between the University of Guelph, the University of Western Ontario, the University of Windsor, and Agriculture and Agri-Food Canada, Génome Québec & Genome Canada SoyaGen.

Author details

Predrag Rajsic[1*], Richard Gray[2], Alfons Weersink[3] and Istvan Rajcan[4]

1 Department of Economics, University of Waterloo, Ontario, Canada

2 Department of Agricultural and Resource Economics, University of Saskatchewan, Saskatoon, Saskatchewan, Canada

3 Department of Food, Agricultural and Resource Economics, University of Guelph, Guelph, Ontario, Canada

4 Department of Plant Agriculture, University of Guelph, Guelph, Ontario, Canada

*Address all correspondence to: prajsic@uwaterloo.ca

References

[1] Fróna D, Szenderák J, Harangi-Rákos M. The challenge of feeding the world. Sustainability. 2019;**11**(20):5816

[2] Valoppi F, Agustin M, Abik F, Morais de Carvalho D, Sithole J, Bhattarai M, et al. Insight on current advances in food science and technology for feeding the world population. Frontiers in Sustainable Food Systems. 2021;**5**:626227

[3] Fuglie KO, Wang SL. New Evidence Points to Robust but Uneven Productivity Growth in Global Agriculture. Global Journal of Emerging Market Economies. 2013;**5**:23-30

[4] OECD. Innovation, Productivity and Sustainability in Food and Agriculture: Main Findings from Country Reviews and Policy Lessons. Paris: OECD Publishing; 2019

[5] Alston JM, Beddow JM, Pardey PG. Agricultural research, productivity, and food prices in the long run. Science. 2009;**325**(5945):1209-1210

[6] Kafatos FC, Eisner T. Unification in the century of biology. Science. 2004; **303**(5662):1257-1257

[7] Warr A, Affara N, Aken B, Beiki H, Bickhart DM, Billis K, et al. An improved pig reference genome sequence to enable pig genetics and genomics research. Gigascience. 2020;**9**(6):giaa051

[8] Love B. Personal communication. (P. Rajsic, ed.). 2013

[9] Ioannidis JP. Differentiating Biases from Genuine Heterogeneity: Distinguishing Artifactual from Substantive Effects. Publication bias in Meta-analysis: Prevention, Assessment and Adjustments. Chichester, West Sussex, England. 2005. pp. 287-302

[10] Open Science Collaboration. Estimating the reproducibility of psychological science. Science. 2015;**349**: 6251

[11] Baker M. Reproducibility crisis. Nature. 2016;**533**(26):353-366

[12] Fanelli D. Opinion: Is science really facing a reproducibility crisis, and do we need it to? Proceedings of the National Academy of Sciences. 2018;**115**(11): 2628-2631

[13] Rasheed A, Hao Y, Xia X, Khan A, Xu Y, Varshney RK, et al. Crop Breeding Chips and Genotyping Platforms: Progress, Challenges, and Perspectives. Molecular Plant. 2017;**10**:1047-1064

[14] Porter, M. E. (2011). Competitive advantage of nations: creating and sustaining superior performance. simon and schuster

[15] Klerkx L, van Mierlo B, Leeuwis C. Evolution of systems approaches to agricultural innovation: concepts, analysis and interventions. In: Darnhofer I, Gibbon D, Dedieu B, editors. Farming Systems Research into the 21st Century: The New Dynamic. Netherlands, Dordrecht: Springer; 2012. pp. 457-483

[16] Spielman DJ, von Grebmer K. Public-private Partnerships in Agricultural Research: An Analysis of Challenges Facing Industry and the Consultative Group on International Agricultural Research. Washington DC: International Food Policy Research Institute (IFPRI); 2004

[17] Heisey PW, Fuglie KO. Public agricultural R&D in high-income countries: Old and new roles in a new funding environment. Global Food Security. 2018;**17**:92-102

[18] Gray RS, Kingwell RS, Galushko V, Katarzyna B. Intellectual Property Rights and Canadian Wheat Breeding for the 21st Century. Canadian Journal of Agricultural Economics/Revue canadienne d'agroeconomie. 2017;**65**:667-691

[19] Varshney RK, Kudapa H, Roorkiwal M, Thudi M, Pandey MK, Saxena RK, et al. Advances in genetics and molecular breeding of three legume crops of semi-arid tropics using next-generation sequencing and high-throughput genotyping technologies. Journal of Biosciences. 2012;**37**:811-820

[20] Galushko V, Gray R. Twenty five years of private wheat breeding in the UK: Lessons for other countries. Science and Public Policy. 2014;**41**:765-779

[21] Morris M, Edmeades G, Pehu E. The global need for plant breeding capacity: what roles for the public and private sectors? HortScience. 2006;**41**:30-39

[22] Oppenheim C. The correlation between citation counts and the 1992 research assessment exercise ratings for British research in genetics, anatomy and archaeology. Journal of Documentation. 1997;**53**:477-487

[23] Li L, Zhang H. Confidentiality and information sharing in supply chain coordination. Management Science. 2008;**54**:1467-1481

[24] Ioannidis JP, Ntzani EE, Trikalinos TA, Contopoulos-Ioannidis DG. Replication validity of genetic association studies. Nature Genetics. 2001;**29**:306-309

[25] Lohmueller KE, Pearce CL, Pike M, Lander ES, Hirschhorn JN. Meta-analysis of genetic association studies supports a contribution of common variants to susceptibility to common disease. Nature Genetics. 2003;**33**:177-182

[26] NCI-NHGRI Working Group on Replication in Association Studies. Replicating genotype–phenotype associations: what constitutes replication of a genotype–phenotype association, and how best can it be achieved? Nature. 2007;**447**:655-660

[27] Hirschhorn JN, Lohmueller K, Byrne E, Hirschhorn K. A comprehensive review of genetic association studies. Genetics in Medicine. 2002;**4**:45-61

[28] Collard BC, Mackill DJ. Marker-assisted selection: an approach for precision plant breeding in the twenty-first century. Philosophical Transactions of the Royal Society, B: Biological Sciences. 2008;**363**:557-572

[29] de Dorlodot S, Forster B, Pagès L, Price A, Tuberosa R, Draye X. Root system architecture: opportunities and constraints for genetic improvement of crops. Trends in Plant Science. 2007;**12**: 474-481

[30] Xu Y, Crouch JH. Marker-assisted selection in plant breeding: from publications to practice. Crop Science. 2008;**48**:391-407

[31] Scheben A, Batley J, Edwards D. Genotyping-by-sequencing approaches to characterize crop genomes: choosing the right tool for the right application. Plant Biotechnology Journal. 2017;**15**: 149-161

[32] Singh K, Ang SH, Leong SM. Increasing Replication for Knowledge Accumulation in Strategy Research. Journal of Management. 2003;**29**: 533-549

[33] Rajsic P, Weersink A, Navabi A, Pauls KP. Economics of genomic selection: the role of prediction accuracy and relative genotyping costs. Euphytica. 2016;**2**:259-276

Chapter 9

Circular Business Models and Global Value Chains: The Case of MagProtein

Natalie Beinisch

Abstract

Global value chain research has focused on the governance of diffuse modes of production and trade that travel in a single direction, however Circular Economy business models are by definition closed economic systems, where it is possible for two parties to act as both buyers and sellers. How then should Circular Business models be understood within the context of international trade systems? Based on an approach which explores the principles that businesses make decisions to set business strategy and the processes through which businesses establish and build external relationships, this chapter examines the case of MagProtein, a company based in Nigeria that produces protein for animal feed and fertilizer using black soldier flies. This chapter finds that although MagProtein's business model centered on local production and distribution, improving operational efficiency and scale depended upon the activities and organizations participating in global value chains.

Keywords: circular economy, global value chains, circular business models, black soldier flies, MagProtein

1. Introduction

Global value chain research focuses on spatially diffuse, uni-directional trade relationships between organizations. However, because Circular Economy approaches are by definition "closed" economic systems that can also blur the lines buyers and sellers, the relationship between global value chains and circular business models is unclear. To date, the global value chain literature has addressed the circular economy as a topic of theoretical interest, with the question of how circular business models are connected to global value chains figuring as a critical topic. Regardless, empirical work that examines the linkages between global value chains and circular business models is limited, given that trade which can be classified as "circular" is very low.

Through the case of MagProtein, a company that produces protein for animal feed and fertilizer using black soldier flies in Nigeria, this chapter explores the principles and processes that link global value chains with circular business models. This chapter begins with a discussion of how the circular economy fits within the global value chain literature and the questions that have been raised about circular business

IntechOpen

models in global value chains. Following a description of the research methodology, this chapter describes why the insect rearing sector is important to the circular economy and the types of business models that have been identified to develop the market for insect protein in developing countries. Next, it describes the business model of MagProtein and the process it went through to identify and integrate multinational companies as key partners in its business. This chapter concludes that even in cases where circular business models are designed for local production and consumption, they are interdependent with global value chains and the organizations that participate and support them.

2. Global value chains and the circular economy

Global Value Chain (GVC) literature emerged as a result of the "globalization" of trade, a phenomenon which describes the "functional integration and coordination of internationally dispersed activities" ([1], p. 41). GVC analysis focuses on mapping and explaining the relationships between organizations that participate in globalized supply chains.

One of the first distinctions made of GVCs is between those that are "buyer-driven" and those that are "producer driven" [2]. The significance of this distinction is that it helps to explain the political and social relationships between participants in the chain. For example, it helps to explain why and how producers comply with environmental and social standards set by "lead" buyers [3, 4], sets expectations about the conditions under which economic and environmental upgrading of the value chain occurs [5, 6] and helps to explain why increases in labour costs do not necessarily lead to changes in value chain organization [7, 8].

GVC analysis was developed to describe and explain the organization of and relationships in spatially diffuse value chains, where purchasing and supply relationships flowed in one direction. By contrast, "circular economy" challenges the idea that trade of materials flow in one direction [9]. Terms such as "industry symbiosis" furthermore describe how waste materials from one industrial process may be converted to other economically productive uses. This has primarily, but not exclusively been observed to take place within shorter geographical distances such as eco-industrial parks, where waste is most efficiently converted into productive energy or material inputs [10]. The "symbiotic" dimension of circular value chains also means that value chain relationships are more complex, as the lines between "buyer" and "supplier" are blurred.

Another distinction of circular economy approaches is on business model innovation that focuses on principles of "dematerialization", "sufficiency" and "efficiency". This calls for significant changes in terms of the way we conceive of producer and consumer relationships. For example, concepts such as "sharing economy" and "product-as-a-service" reconfigure ownership and purchasing structures [11–13] and the emphasis on business model innovation as a means to achieve circular economy goals increases the importance of business eco-system development [14]. This involves different organizational relationships than the producer networks described by Gereffi [1] in sectors such as garment and toy production.

It is the dramatic changes in the definition of value chain relationships that make Circular economy approaches appealing from a "Global South" perspective. "Closed loop" concepts favor local economic development vis-a-vis spatially distributed chains, creating more potential to develop and retain more value-add activities in developing countries. The emphasis of circular economy business models on knowledge retention, digitization and innovation reinforce this point of view and is the rationale by which organizations such as the African Development Bank have set up Circular Economy programmes. Some authors, such as Wellesley et al. [15] also point out that developing countries are inherently more circular in terms of consumption behaviors and so can also act as models for developed countries. Furthermore, the priority of effective waste management to enable circular processes creates a space for the interests and voices of vulnerable economic participants such as waste pickers and the informal sector to be represented [16–18]. This helps to establish Circular Economy principles as ones that are intentionally more egalitarian and socially inclusive compared to alternative trade models.

As a proportion of material production, circular activities are relatively limited [19]. By the estimate of the consultancy Circle Economy, which calculates the volume of waste produced and subtracts from the volume of materials recovered, circular economic activities represent about 9% of total economic activity globally [20]. Thus, with limited empirical cases available, the exploration of circular value chains through a GVC approach has been primarily focused on developing research agendas and theoretical pathways to shape our understanding of the ways that the circular economy approaches influence value chain relationships [21, 22]. The question of how organizational forms and business models are suited to the circular economy on a global scale is one among a number of questions connected to the emerging research agenda ([22], p. 24). By exploring the case of MagProtein, a Nigeria-based company that produces animal protein and fertilizer using black soldier flies, which sought to develop partnerships with leading multinational food manufacturers operation in Nigeria, this chapter helps to develop an empirical foundation to help us to understand the linkages between circular economy business models and the global economy.

3. Methodology

This chapter uses a single case approach to examine the question of how business models are suited to the circular economy on a global scale. According to Yin [23], there are three types of case-based research, which include 1) Exploratory, that is intended to develop insights into relatively unknown subjects; 2) Descriptive, deployed to describe the function of known phenomenon and 3) Explanatory, which is used to establish cause and effect between variables. Having selected a case that has established linkages with a multinational company, the case can be classified as descriptive insofar as it seeks to describe how these linkages developed but it does not seek to establish cause and effect. Given there is a diversity of business models that can be classified as "circular" and each have different structural characteristics, it would be presumptuous to make generalizations from this case about circular business models and their relationship to global value chains more broadly.

The descriptive approach is structured around two assumptions. Firstly, that organizations pursue opportunities based on a logic or "knowledge structures" [24]; that is their decisions are driven by principles which shape their external relationships. The second assumption is that cross-organizational relationships are developed through iterative steps and these relationships are dynamic.[1] In other words, a process

[1] The iterative process through which inter-organizational relationships are formed is documented in Beinisch [25].

of interactions also governs how organizational relationships form and grow. These assumptions, that organizational principles drive decision-making, and that interactive processes shape how those decisions are implemented and change are the dimensions that are used to examine how MagProtein's business model evolved to develop linkages within a global value chain.

Data collection centered on the case itself. One telephone meeting and three face-to-face meetings were held with MagProtein. Face-to-face meeting locations were at the administrative office of MagProtein in Ikorudu, Nigeria at their facilities, in Epe, Nigeria and at a third location in Lagos, Nigeria. Further online interviews were carried out with black soldier fly eco-system organizers, based in Ibadan, Nigeria and Groningen, the Netherlands and with an international beverage company that is a strategic partner of MagProtein. While the sample size of interviews is small, comprising 5 people interviewed between 2 and 4 times, the primary intention of the case is to map business decision-making principles, processes and interactions, key of which was between MagProtein and its international partners. Interviews beyond the scope of this interaction would not have yielded insights relevant to the case. The saturation principle was used throughout the interview process to assure that there was adequate data to support themes emerging from the case and to identify conflicts in the information gathered.

Public data about the black soldier fly, animal protein and fertilizer sectors in Nigeria is limited. Interviewees were asked to share available documents about these sectors, which were used to triangulate the study. Topics raised in the interviews were likewise cross-referenced with publicly available materials. Where figures are referenced in this study, they come from publicly available sources.

4. Business models for insect protein in Africa

Black soldier flies are primarily reared to produce protein for animal feed. Biowaste from insect production, called frass, can be sold as organic fertilizer. From an environmental perspective, black solider flies are seen as a promising protein alternative because organic waste can be used as a substrate to grow flies in a hygienic way [26] and the space needed to rear flies is marginal compared to other types of protein. As such, insect protein production can address issues such as carbon emissions from food wastage, food security and land use concomitantly [27–30]. As Soma [31] points out in her article advocating for the development of black soldier fly protein sector in Africa, the fact that larvae feed on a wide range of organic waste materials, the waste that they produce in their lifecycle can be applied back to agricultural activity in a hygienic and safe way means that black soldier fly rearing is a critical part of a "circular animal system, where the total protein supply is increased without the need for additional resources" ([31], p 59).

The potential market for insect-based protein across Africa is argued to be significant. According to a report by Manufacturing Africa, a Foreign Commonwealth Development Office (FCDO) programme, the growth potential for insect-protein in Africa is substantial, with the report authors estimating annual revenue generation of 500–700 million USD for the sector by 2030. This estimate is noteworthy because insect-based protein currently represents a tiny fraction the total market for animal feed protein in Africa, with very few companies across the continent producing above 100 kilograms of protein per day. The argument that the market is ripe for substantial growth stems from a view that insect-based proteins are highly cost-effective alternatives imported animal feed [32].

The market for insect-based protein in Nigeria is only beginning to take shape. While research on the viability of black soldier flies for fish meal were carried out as early as 2000 [33], it was not until 2016 that actions were taken to develop the sector, through the establishment of the Insect4Feed Cluster, a Dutch-funded programme that was set up to help reduce the costs of feed to farmers. Without much interest from international feed companies to invest in Nigeria, the Insect4Feed Cluster decided to focus on developing opportunities for small-hold farmers and young entrepreneurs in addition to supporting the development of technical standards (Interview with Technical Expert December 2022).

While insect protein production is highly technical, with capital investment needs at the industrial scale [34], it is possible to deploy business models that are suitable for micro and small-scale producers. For example, one technical expert interviewed likened setting up a black-soldier fly rearing business to purchasing furniture at IKEA insofar as there is standardized set of equipment and relatively low capital expenditure needed for a small producer to set up an insect rearing operation (Interview with Technical Expert November 2022).

At the small-holder level, three types of business models that reduce feed input costs for small-scale farmers have been identified by Soma [31] and Chia et al. [35]. This includes models where:

1. Existing local feed companies incorporate insect protein as a feed ingredient, cutting the costs of feed inputs for farmers;

2. Farmers cooperate with local feed companies by producing and selling larvae to feed companies, with the feed being sold back to farmers;

3. Farm-based production of larvae and feed.

At the farm-production level, technologies such as feeding apps help to create "recipes" for farmers to combine with other accessible ingredients to create feed that is the optimal nutritional value (Interview with Technical Expert December 2022).

Still, while many aspects of black soldier fly rearing make it accessible to small producers and farmers, it is very challenging to produce insect protein at an industrial scale in Nigeria because it is technically difficult to maintain a stable insect population and to manage its lifecycle. Infrastructure issues such as unstable electricity make the work of lifecycle management challenging. It is nevertheless industrial scale production of black soldier flies at the local level that will make cheaper protein alternatives more widely available to farmers in Nigeria.

5. MagProtein: global value chains and local production and consumption

MagProtein is among the largest producers of insect protein in Africa and is the largest producer in Africa, with production output of 800 kg per day and capacity of 1.2 tons per days in 2022.

The business was established in 2017 by George Thorpe and Feyi Akinlotan. Having been trained as an accountant and working in the financial services industry for several years, Thorpe was focused on setting up his own enterprise. Initially focused on a cold chain operations, a chance meeting between Akinlotan, who was developing a business plan for black soldier flying rearing, and Thorpe's father, led Akinlotan and Thorpe to partner.

Based on research in the aquaculture sector, Akinlotan and Thorpe saw an opportunity to provide lower-cost inputs to fish feed. With a view that there was a market need for feed products and the fact that Akinlotan and Thorpe believed insects were in abundance across Africa, the team was convinced there was a strong business case and so decided to pilot production.

The goal of the pilot was to determine whether the duo could produce high quality feed for fish from black soldier flies. The piloting stage was highly experimental. Knowledge about black soldier fly rearing was mainly collected through online research and trial and error. In fact, at the time there was no black soldier larvae available in the Nigerian market, so Akinlotan traveled by bus to Benin to acquire the team's first sample. Akinlotan and Thorpe also experimented with variables such as lighting, mass balance, and temperature. For example, it was by accident that the team found that larvae reacted to light and through experience that the team found that even very slight changes in temperature affected the insect lifecycle.

Results from their early pilot produced 100% insect based feed with for catfish a growth performance of 75% compared to traditional feed, which the MagProtein team saw as very promising, given that additional ingredients could be added to the insect meal at a relatively low cost to achieve comparable growth yields.

However, a problem the team had to contend with was the stability of the yields of the insects they were rearing. MagProtein's initial business plan had set a target of one ton of feed production per day. This was based on the investment they were making into the facility and their assumptions about running it, but did not take into account the technical and logistical issues associated with acquiring substrate for the insects and managing their lifecycle. Accordingly, the production yields of MagProtein were one tenth the order of magnitude of their projections in 2018.

Falling short of their targets, the MagProtein team searched for help from technical experts outside of Nigeria. The first set of experts MagProtein consulted was Jeffrey Tomberlin, a Professor of Agri-life research at Texas A&M University. According to Thorpe, in meeting Tomberlin, "we realized there are many things we don't understand; that there were problems in the breeding and we didn't know how to control factors and keep them stable" (Interview with MagProtein December 2022). Tomberlin referred the MagProtein Team to the EVO Consortium, a research consultancy that specialized in black soldier fly rearing. Working with the Consortium, MagProtein set up and implemented technical specifications and processes for its production that were designed to produce the optimal yield of insect larvae. While the MagProtein team found the technical support was an important developmental step, "it was not enough to take us to the next level. We needed the technology, to install equipment and to set up the production flow" (Interview with MagProtein December 2022) and so the team began to search for alternative operational support.

In parallel to Akinlotan and Thorpe's work to improve the technical aspects of their operations, other developments were taking place which led the team to reconsider their business model. The first development was related to how the team was procuring organic waste. Initially, MagProtein purchased market waste, however this was logistically complex, expensive and the moisture content of the waste was too high to be used without additional processing. MagProtein subsequently turned to procure spent grain from traders. However, MagProtein found that the quality of deliveries was inconsistent. This was due to the fact that traders sold grain that was mixed with other materials or they did not deliver the grain at all as MagProtein competed with companies who used the grain to produce animal feed. Furthermore, spent grain needed to be handled in specialized storage facilities so that it would not go bad, which MagProtein did not have the capacity to handle.

The difficulties of sourcing a consistent supply of organic waste led MagProtein to determine that it would be more efficient to work directly with industrial producers of organic waste such as food and beverage manufacturing companies.

The second development was that an encounter with a local farmer led MagProtein to turn insect excretion, called frass, into another product category. While the team was aware that frass could be converted to fertilizer and, based on the advice of the EVO Consortium, were allowing frass to stockpile, they did not invest in frass production until they met a local farmer who came to their facility for a visit. MagProtein agreed to provide a couple of bags of the frass at the end of his visit, with the farmer returning a short time later to procure a "boatload" (Interview with MagProtein November 2022). Further research undertaken by the team helped them to understand the commercial potential of frass, which was a high potential fertilizer that was competitive with imported products.

The conclusion that industrial waste producers were the optimal business partners to MagProtein and that there was commercial value to frass helped MagProtein's founders to redefine its business model towards 2020. Importantly, they determined that by working closely with large manufacturers, they could optimize the management of their own supply chain and help large manufacturers with their waste streams. They also concluded that in addition to supplying waste, corporate partners could also act as clients, either by purchasing insect protein as a feed ingredient or by purchasing fertilizer for food growers that were supplying manufacturers. Based on these revelations, the company redefined its business plan to seek industrial scale manufacturers as both suppliers and clients.

Understanding that they also had to improve facility operations to expand breeding capacity, MagProtein sent an email to Avaro Manzanares Rosenberg, who ran a Dutch-based business called Insect Systems. To the surprise of MagProtein, Manzanares was "enthused by the big ambition" of MagProtein (Interview with MagProtein December 2022) and decided to visit Nigeria within two weeks of their first correspondence, arriving in early 2020. Manzanares was impressed by MagProtein's production volumes and introduced the founders to New Generation Nutrition, a Netherlands-based insect protein specialist that was leading the development of the Insect4Feed Cluster in Nigeria.

The introduction came at an opportune time for MagProtein as the company had been unsuccessfully trying to get in touch with an international food manufacturer to find a reliable supply of spent grain and New Generation Nutrition had direct links to the organization and so facilitated a direct meeting.

The multi-national company had set global "zero waste-to landfill" reduction targets for its factories and production sites and it was the responsibility of local organizations to determine how to meet these targets. Among the waste streams produced by the company were spent grain and spent yeast. While the applications for spent grain were more developed; in addition to insect-rearing; spent grain could be used as a direct input into animal feed and so the company had a productive way to valorize spent grain, the company believed that it's disposal of spent yeast needed to be optimized. At the time of meeting the MagProtein team in early 2022, the company was delivering its spent yeast to a composting company. However, composting yeast into fertilizer was time consuming and, in the view of the company, the composting process produced unnecessary greenhouse gasses. Due to the corrosive nature of spent yeast, the alternatives to convert spent yeast into other products were more limited. Based on the interests of the company, to minimize the environmental impact and maximize the valorization of all its waste streams, and the interests of MagProtein, to identify a stable supply of feedstock for its insects, the company recommended to trial delivery of spent yeast, a proposal which was accepted by MagProtein.

The results of the pilot were positive, with the initial pilot volumes expanding from one thousand to twenty thousand liters of spent yeast per day. Because spent yeast has a high proportion of water content, which larvae also need for growth and reproduction, MagProtein was able to reduce water and energy expenditures. The reductions of input costs meant that the insect protein produced by Magprotein was approximately 50% cheaper that fishmeal on absolute terms and 15–20% cheaper based on its protein value, making it a highly competitive feed input. Secondly, from the perspective of the multinational company, the application of spent yeast to rear black soldier flies was more environmentally efficient and logistically straightforward compared to composting.

The company's interests to maximize the operational and environmental efficiency of their supply chains also meant the company was open to explore how the fertilizer produced by MagProtein could be distributed through the company's outgrower schemes. According to the company, this level of efficiency was attractive because "one process is being used to create two resources" (Company Interview December 2022). At the time of writing, yield trials were underway, serving as validation that the strategy it had developed from its production ambitions, its early experience procuring organic waste streams and its decision to produce fertilizer, was an achievable growth path.

6. Discussion: principles, processes and the integration of circular business models into global value chains

The chapter explores how circular business models are linked to global value chains, through the case of MagProtein. Because circular business models have several characteristics that contradict our assumptions about the structure of global value chains, specifically that trade is more likely to take place in "closed loops" and parties in a trade relationship may take on the roles of both buyers and sellers, it is valuable to interrogate whether linkages between circular business models and global value chains exist, as well as the structure of these linkages.

The chapter has selected the case of MagProtein because insect rearing is a quintessentially "circular" business insofar as it produces valuable forms of energy from organic waste, using a very small environmental footprint. While possible business models have been identified that are possible to deploy at the small-holder level in Nigeria, MagProtein pursued a business model that was designed to be industrial in scale. The implications of this were that MagProtein needed to work with industrial-scale partners to secure stable volumes of waste resources.

Potential industrial scale partners were international companies because it was international companies that were producing the volumes of waste that MagProtein needed. International companies also had globally established targets that centered on zero-waste manufacturing principles, which made them more likely to be open to partnering with MagProtein. Thus, while MagProtein's entry into the "global value chain" of a multinational food manufacturer was very local because production and distribution it's products remained within Nigeria, it was the presence and activity of industrial scale production that enabled the business model of MagProtein to materialize.

Another aspect of the development of MagProtein's business model that is notable is the extent to which global value chains are not just global in relation to trade but also in terms of research and development and market access.

International linkages supported the development and growth of MagProtein's business. Outreach to international organizations was an essential part of the company's early business development in terms of technical, operational and market linkages. In this respect, the study of the integration of circular business models into global value chains, especially those which emerge in developing countries, should consider the role played by international technical experts and the ways that they support circular businesses to integrate into global value chains.

Author details

Natalie Beinisch
Circular Economy Innovation Partnership, Nigeria

*Address all correspondence to: natalie@ceipafrica.org

References

[1] Gereffi G. International trade and industrial upgrading in the apparel commodity chain. Journal of International Economics. 1999;**48**(1):37-70

[2] Gereffi G. The organization of buyer-driven global commodity chains: How US retailers shape overseas production networks. Commodity chains and global capitalism. 1994:95-122

[3] Konefal J, Mascarenhas M, Hatanaka M. Governance in the global agro-food system: Backlighting the role of transnational supermarket chains. Agriculture and Human Values. 2005;**22**(3):291-302

[4] Ponte S, Gibbon P. Quality standards, conventions and the governance of global value chains. Economy and Society. 2005;**34**(1):1-31

[5] Valentina DM et al. Environmental upgrading in global value chains. In: Handbook on Global Value Chains. Northampton: Edward Elgar Publishing; 2019

[6] Gereffi G. Economic upgrading in global value chains. In: Handbook on Global Value Chains. Northampton: Edward Elgar Publishing; 2019

[7] Valentina DM, Di Maria E, Micelli S. Environmental strategies, upgrading and competitive advantage in global value chains. Business Strategy and the Environment. 2013;**22**(1):62-72

[8] Humphrey J, Schmitz H. Governance and Upgrading: Linking Industrial Cluster and Global Value Chain Research. Vol. 120. Brighton: Institute of Development Studies; 2000

[9] Wang E, Miao C, Chen X. Circular economy and the changing geography of international trade in plastic waste. International Journal of Environmental Research and Public Health. 2022;**19**(22):15020

[10] Neves A et al. A comprehensive review of industrial symbiosis. Journal of Cleaner Production. 2020;**247**:119113

[11] Centobelli P et al. Designing business models in circular economy: A systematic literature review and research agenda. Business Strategy and the Environment. 2020;**29**(4):1734-1749

[12] Sposato P et al. Sharing economy and circular economy. How technology and collaborative consumption innovations boost closing the loop strategies. Environmental Engineering & Management Journal (EEMJ). 2017;**16**(8):1797-1806

[13] Schwanholz J, Leipold S. Sharing for a circular economy? An analysis of digital sharing platforms' principles and business models. Journal of Cleaner Production. 2020;**269**:122327

[14] Santa-Maria T, Vermeulen WJV, Baumgartner RJ. How do incumbent firms innovate their business models for the circular economy? Identifying micro-foundations of dynamic capabilities. Business Strategy and the Environment. 2022;**31**(4):1308-1333

[15] Preston F, Lehne J, Wellesley L. An inclusive circular economy. Priorities for Developing Countries. 2019

[16] Gutberlet J et al. Waste picker organizations and their contribution to the circular economy: Two case studies from a global south perspective. Resources. 2017;**6**(4):52

[17] Gutberlet J, Carenzo S. "waste pickers at the heart of the circular economy: A perspective of inclusive recycling from the global south." worldwide waste. Journal of Interdisciplinary Studies. 2020;**3**:1

[18] Velis C. Waste pickers in global south: Informal recycling sector in a circular economy era. Waste Management & Research. 2017;**35**(4):329-331

[19] Aguilar-Hernandez GA et al. The circularity gap of nations: A multiregional analysis of waste generation, recovery, and stock depletion in 2011. Resources, Conservation and Recycling. 2019;**151**:104452

[20] Circle Economy. Circularity Gap Report, 2021. Available from: CGR 2021 (circularity-gap.world)

[21] Awan U et al. Industry 4.0 and circular economy in an era of global value chains: What have we learned and what is still to be explored? Journal of Cleaner Production. 2022;**371**:133621

[22] Hofstetter JS et al. From sustainable global value chains to circular economy—Different silos, different perspectives, but many opportunities to build bridges. Circular Economy and Sustainability. 2021;**1**(1):21-47

[23] Yin RK. Case Study Research: Design and Methods. Vol. 5. New York: Sage; 2009

[24] Oliver D, Roos J. Decision-making in high-velocity environments: The importance of guiding principles. Organization Studies. 2005;**26**(6):889-913

[25] Beinisch N. Making it Work: The Development and Evolution of Transnational Labour Regulation [Thesis]. London, England: London School of Economics and Political Science; 2017

[26] Henry M et al. Review on the use of insects in the diet of farmed fish: Past and future. Animal Feed Science and Technology. 2015;**203**:1-22

[27] Jagtap S et al. Codesign of food system and circular economy approaches for the development of livestock feeds from insect larvae. Food. 2021;**10**(8):1701

[28] Liceaga AM. Processing insects for use in the food and feed industry. Current opinion in insect science. 2021;**48**:32-36

[29] Madau FA et al. Insect farming for feed and food production from a circular business model perspective. Sustainability. 2020;**12**(13):5418

[30] van Huis A. Potential of insects as food and feed in assuring food security. Annual Review of Entomology. 2013;**58**:563-583

[31] Soma K. Ensuring accessibility of insect-based feed to small-scale fish farmers in Kenya and Uganda. In: A Journey into the world's Food Systems in Search of Losses, Waste and Ways to Solve them. Vol. 58. Wegeningen: Wegeningen University & Research; 2022

[32] Bouchene L, et al. Africa's green manufacturing crossroads: Choices for a low-carbon industrial future. McKinsey. 2021. Available from: https://www.mckinsey. com/business-functions/sustainability/our-insights/africas-green-manufacturing-crossroads-choices-for-a-low-carbon-industrial-future

[33] Oluokun JA. "upgrading the nutritive value of full-fat soyabeans meal for broiler production with either fishmeal or black soldier fly larvae meal (Hermetia illucens)." Nigerian. Journal of Animal Science. 2000;**3**:2

[34] Lorrette B, Sanchez L. New lipid sources in the insect industry, regulatory aspects and applications. OCL. 2022;**29**:22

[35] Chia SY et al. Insects for sustainable animal feed: Inclusive business models involving smallholder farmers. Current Opinion in Environmental Sustainability. 2019;**41**:23-30